CAMBRIDGE LIBRARY COLLECTION

Books of enduring scholarly value

Zoology

Until the nineteenth century, the investigation of natural phenomena, plants and animals was considered either the preserve of elite scholars or a pastime for the leisured upper classes. As increasing academic rigour and systematisation was brought to the study of 'natural history', its sub-disciplines were adopted into university curricula, and learned societies (such as the London Zoological Society, founded in 1826) were established to support research in these areas. These developments are reflected in the books reissued in this series, which describe the anatomy and characteristics of animals ranging from invertebrates to polar bears, fish to birds, in habitats from Arctic North America to the tropical forests of Malaysia. By the middle of the nineteenth century, this work and developments in research on fossils had resulted in the formulation of the theory of evolution.

The Humble-Bee

This classic work by F.W.L. Sladen (1876–1921) was published in 1912. Captivated by the bees in the grounds of his home, he produced his first essay on them in 1892, and later began to keep bees and produce honey as a livelihood, at the same time studying and breeding honeybees. Moving in 1912 to Canada, he eventually received the title of Dominion Apiarist, but unfortunately his work was cut short by his sudden death in 1921. When this book was published, there were no detailed accounts of the life cycles of the humble-bee (or bumblebee) species. Sladen provides these, with a guide to distinguishing the different British species (with colour plates which can be viewed at http://www.cambridge.org/9781108075725), and instructions on how to domesticate these important pollinators. Combining the enthusiasm of a naturalist with the precision of a scientist, this work is of continuing interest and importance in environmental studies.

Cambridge University Press has long been a pioneer in the reissuing of out-of-print titles from its own backlist, producing digital reprints of books that are still sought after by scholars and students but could not be reprinted economically using traditional technology. The Cambridge Library Collection extends this activity to a wider range of books which are still of importance to researchers and professionals, either for the source material they contain, or as landmarks in the history of their academic discipline.

Drawing from the world-renowned collections in the Cambridge University Library and other partner libraries, and guided by the advice of experts in each subject area, Cambridge University Press is using state-of-the-art scanning machines in its own Printing House to capture the content of each book selected for inclusion. The files are processed to give a consistently clear, crisp image, and the books finished to the high quality standard for which the Press is recognised around the world. The latest print-on-demand technology ensures that the books will remain available indefinitely, and that orders for single or multiple copies can quickly be supplied.

The Cambridge Library Collection brings back to life books of enduring scholarly value (including out-of-copyright works originally issued by other publishers) across a wide range of disciplines in the humanities and social sciences and in science and technology.

The Humble-Bee

Its Life-History and How to Domesticate It,
with Descriptions of All the
British Species of Bombus and Psithyrus

F.W.L. SLADEN

CAMBRIDGE
UNIVERSITY PRESS

CAMBRIDGE
UNIVERSITY PRESS

University Printing House, Cambridge, CB2 8BS, United Kingdom

Cambridge University Press is part of the University of Cambridge.

It furthers the University's mission by disseminating knowledge in the pursuit of
education, learning and research at the highest international levels of excellence.

www.cambridge.org
Information on this title: www.cambridge.org/9781108075725

© in this compilation Cambridge University Press 2014

This edition first published 1912
This digitally printed version 2014

ISBN 978-1-108-07572-5 Paperback

THE HUMBLE-BEE

MACMILLAN AND CO. LIMITED
LONDON · BOMBAY · CALCUTTA
MELBOURNE

THE MACMILLAN COMPANY
NEW YORK · BOSTON · CHICAGO
DALLAS · SAN FRANCISCO

THE MACMILLAN CO. OF CANADA, LTD.
TORONTO

A PET QUEEN OF *BOMBUS TERRESTRIS* INCUBATING HER BROOD.

(See page 139.)

THE HUMBLE-BEE

ITS LIFE-HISTORY
AND HOW TO DOMESTICATE IT

WITH

DESCRIPTIONS OF ALL THE BRITISH SPECIES
OF *BOMBUS* AND *PSITHYRUS*

BY

F. W. L. SLADEN

FELLOW OF THE ENTOMOLOGICAL SOCIETY OF LONDON
AUTHOR OF 'QUEEN-REARING IN ENGLAND'

ILLUSTRATED WITH PHOTOGRAPHS AND DRAWINGS BY THE AUTHOR
AND FIVE COLOURED PLATES PHOTOGRAPHED DIRECT
FROM NATURE

MACMILLAN AND CO., LIMITED
ST. MARTIN'S STREET, LONDON
1912

PREFACE

THE title, scheme, and some of the contents of this book are borrowed from a little treatise printed on a stencil copying apparatus in August 1892.

The boyish effort brought me several naturalist friends who encouraged me to pursue further the study of these intelligent and useful insects. Of these friends, I feel especially indebted to the late Edward Saunders, F.R.S., author of *The Hymenoptera Aculeata of the British Islands*, and to the late Mrs. Brightwen, the gentle writer of *Wild Nature Won by Kindness*, and other charming studies of pet animals.

The general outline of the life-history of the humble-bee is, of course, well known, but few observers have taken the trouble to investigate the details. Even Hoffer's extensive monograph, *Die Hummeln Steiermarks*, published in 1882 and 1883, makes no mention of many remarkable particulars that I have witnessed, and there can be no doubt that further investigations will reveal more.

An article entitled "*Bombi* in Captivity and
Habits of *Psithyrus*" that appeared in the *Entomologist's Monthly Magazine* for October 1899, contained my original division of the species of *Bombus*
into pollen-storers and pocket-makers, and gave
accounts of the self-parasitism of certain species, and
of the parasitism of *B. terrestris* upon *B. lucorum*.

Among matter now published for the first time
are particulars of the Sladen wooden cover for
artificial nests, and details of my humble-bee house.
By the employment of the covers anybody may
attract humble-bees to nest in his garden, and by
obtaining a little wooden house and furnishing it
as directed, one may study at leisure and in comfort
all the details of their interesting and intelligent
ways.

The study of humble-bees has hitherto been
hampered by the difficulty, encountered even by
experienced entomologists, in separating some of
the species. It is hoped that the colour-photographs,
used in conjunction with the descriptions given,
will now remove this difficulty.

My thanks are due to Messrs. L. S. Crawshaw,
J. W. Cunningham, Geo. Ellison, P. E. Freke,
A. H. Hamm, W. H. Harwood, Rev. W. F.
Johnson, Rev. F. D. Morice, Messrs. E. B.
Nevinson, H. L. Orr, and Rupert Stenton for
information kindly supplied about the distribution

of the rarer species and varieties in different parts of the kingdom, and, in many cases, for the gift or loan of specimens.

I desire also to acknowledge the kind help of the officials of the British Museum and of the Irish National Museum in granting me every facility for examining specimens in these national collections.

F. W. L. SLADEN.

RIPPLE, DOVER,
Feb. 5, 1912.

CONTENTS

ix

ILLUSTRATIONS

xi

PLATES

All the specimens shown in the plates are British, except the worker of *Bombus soroënsis*, which is from the Caucasus Mountains, and differs from British specimens in having the yellow on the abdomen extending farther on the 1st segment.

A millimetre scale is shown on page 154.

I

INTRODUCTION

EVERYBODY knows the burly, good-natured bumble-bee. Clothed in her lovely coat of fur, she is the life of the gay garden as well as of the modestly blooming wayside as she eagerly hums from flower to flower, diligently collecting nectar and pollen from the break to the close of day. Her methodical movements indicate the busy life she leads—a life as wonderful and interesting in many of its details as that of the honey-bee, about which so much has been written. Her load completed, she speeds away to her home. Here, in midsummer, dwells a populous and thriving colony of humble-bees. The details of the way in which this busy community came into being, what sort of edifice the inhabitants have built, how they carry out their duties, and what eventually will become of them will be explained later : it is enough at present to note that the colony, like a hive of honey-bees, consists chiefly of workers, small modified females, whose function in life is not to give birth but to labour for the establishment, bringing home and depositing in cells load after

B

load of sweets, their only relaxation from this arduous toil being domestic work, such as tending the young, building the comb, and keeping the nest clean and tidy.

The supposition that the humble-bee worker is less industrious than the honey-bee is erroneous : she labours with the same zeal and tireless energy, never ceasing until, worn out, she fails one day to return home, and, becoming drowsy and senseless, passes out of existence in the cold of the succeeding night. It is true that in a colony of humble-bees the workers are not nearly so numerous as in a bee-hive, but it is some compensation that they are of a larger size than honey-bees, that they begin field - work at an earlier age, that their hours of labour are longer, commencing earlier in the morning and continuing until later at night, and that they are more hardy, minding less the spells of wind and rain, cloud and cold from which no English spring or summer is free.

One gets a good idea of the ceaseless industry of a colony of humble-bees by watching for a while the mouth of the hole leading to the domicile. Though the total population may not exceed one or two hundred, a minute seldom goes by without several departures and arrivals, and two bees will often return together or pass one another : almost all the returning bees have their hind legs laden with pollen and their abdomens distended with nectar.

Fanciful writers have likened a colony of bees to a kingdom or city : in reality it is an ordinary family, although a large one. There is the mother, whom we call the queen; and who lays the eggs. Her daughters, the workers, do not become independent as soon as they are old enough to be useful, but, as has been remarked, devote their energies to supporting the family and rearing their younger brothers and sisters. One of the peculiarities of the bee family is that all the work is done by the female members. The father has died long before his children are born. The sons are idle, contributing nothing to the stores of the colony; in the honey-bees' family they are maintained entirely at the expense of the colony, and, when food grows scarce, they are turned out to die, but the humble-bee drones maintain themselves, quietly taking their departure from the nest as soon as they are able to fly.

By far the most interesting individual in the humble-bee family is the queen, because of the very eventful life she leads. At first her duties include those of the workers, her brood depending upon her for everything—food, warmth, and protection from enemies. She nurses it with as much motherly devotion, industry, and patience as we see displayed by many birds and mammals in the care of their young: she thus shows much greater capacity and higher intelligence than the queen honey-bee, who, fed and attended by workers throughout her life, is not only incapable of providing for herself, but pays no attention whatever to her offspring, and is merely a machine for laying eggs in enormous numbers.

The humble-bee and the honey-bee are the only

bees in the temperate zone that produce workers and dwell in communities.

The true humble-bees comprise the genus **Bombus.**[1] Seventeen different species of them are found in the British Isles. In addition, there are six British species of the genus **Psithyrus,**[2] comprising the parasitic humble-bees.

Most of the British species of humble-bees are black with bright yellow bands, which, however, are sometimes absent, and with a white, orange, or. red tail. The remaining species are more or less yellow or tawny.

Humble-bees are essentially inhabitants of the north, and they flourish best at about the latitude of Britain. Europe, Central Asia, and North America are well populated with them, especially the mountainous regions. Even in Greenland, Alaska, and other dreary tracts in the far north, where the summer is too short for the existence of honey-bees, a few species are to be found, working diligently during the light nights. "Others," in the words of Shuckard, "occur far away to the north of east, booming through the desolate wilds of Kamtchatka, having been found at Sitka, and their cheerful hum is heard within the Arctic Circle as high as Boothia Felix, thus more northerly than the seventieth

---

[1] Greek βόμβος (Latin *bombus*), humming, buzzing. Dr. Feltoe has kindly called my attention to an interesting passage in Theokritos (*Idyll* iii. 12 ff.): "Would I were a humming-bee (βομβεῦσα μέλισσα), and could enter thy cave, penetrating the ivy and the fern under which thou dost conceal thyself." But βομβυλιός was the word usually employed for the humble-bee, *e.g.* by Aristophanes, *Wasps*, 107, and by Aristotle, *Hist. Anim.* ix. 40 and 43.

[2] Pronounced *psíthïrus.* From Greek ψίθυρος, whispering, twittering, perhaps in allusion to their softer hum.

parallel. They may perhaps with their music often convey to the broken-hearted and lonely exile in Siberia the momentarily cheering reminiscence of joyful youth, and by this bright and brief interruption break the monotonous and painful dulness of his existence, recalling the happier days of yore."[1]

In the Himalayas they are to be met with at all altitudes from 2,000 to over 12,000 feet. But in the plains of India there are none, nor do they exist in Africa, except along the north coast, and Australia and New Zealand have no native species. Where they occur in the tropics they are generally confined to the mountains, although Brazil has a few indolent-looking species.

It is safe to say that the total number of species, i.e. of forms that do not interbreed, exceeds a hundred, and that the lesser varieties amount to more than a thousand.

It is charming to watch a populous colony of humble-bees busy on its comb, each individual wearing the beautiful livery of its particular species. Each species has its own peculiarities of habit and disposition, so that even in the British fauna there are plenty of different natures to study.

Investigating the habits of humble-bees, and experimenting in different ways with them, has been a source of great pleasure to me since boyhood. Colonies have been kept under observation in artificial domiciles ; the ins and outs of their lives have

[1] *British Bees*, by W. E. Shuckard, 1868, page 78.

thus been laid bare, and their requirements, which are quite different from those of the honey-bee, have been studied and supplied. In addition, various attempts have been made with queens to establish the colonies artificially; these have been partially successful, and have revealed several interesting facts about humble-bees, especially respecting their adaptability to treatment and their intelligence.

Only a very few of the numerous queens that set out in the spring with so much promise succeed in establishing colonies. Their failure is due not so much to unfavourable weather as to the attacks of enemies. In its early stages the brood is very liable to be eaten by ants or mice : when this danger is past a humble-bee of the idle genus *Psithyrus* may enter the nest, kill the queen, and make slaves of her children; at a still later period the brood may be consumed by the caterpillars of a wax-moth. As soon as any of these foes have found and entered the nest there is no escape for the inhabitants from destruction, and it has given me a good deal of pleasure to try and protect the bees that have been under my care from them.

It may be asked : Can humble-bees be made to produce honey for human consumption ? Under favourable conditions humble-bees store honey, the flavour of which, as most schoolboys know, is excellent; but, unfortunately, the amount in each nest never exceeds a few ounces, so that to obtain a quantity many colonies would have to be kept, and even then the work of collecting it would be laborious.

The tongue of the humble-bee is much longer than that of the honey-bee, consequently she can extract honey from flowers having long narrow tubes, such as red clover, honeysuckle, and horehound, which are seldom or never visited by honey-bees. As a rule these flowers are very melliferous. Indeed, the heads of the red clover contain more honey than almost any other flower, a fact appreciated by children, who pull out the tubes and suck them. Humble-bees have almost a monopoly of the vast amount of honey that is produced in a red clover field, but there are not enough of them to gather much of it.

Nevertheless humble-bees are extremely valuable for fertilising the numerous flowers that they frequent. Whole groups of plants bearing long-tubed flowers, including many species valuable to man, depend chiefly upon humble-bees for their propagation. Charles Darwin, in the *Origin of Species*, said : " I find from experiment that humble-bees are almost indispensable to the fertilisation of the heartsease (*Viola tricolor*), for other bees do not visit this flower." In consequence of the absence of humble-bees in New Zealand it was found that the red clover did not produce seed freely. So in November and December 1884 a number of queens were sent from England to that country, with the result that two species, *B. terrestris* and *B. ruderatus*, have become established there, and

the red clover now yields a plentiful crop of seed.
Unfortunately, *B. terrestris* has a trick of biting
holes in such flowers as the broad-bean, snapdragon,
and foxglove, close to the honey-glands, to abstract
the honey. This, in New Zealand, has resulted in
damage to the seed-vessels of certain flowers, and
the seed-growers there would now be glad to have
this species supplanted by another.

The world, which to us consists of sights and
sounds, to the humble-bees is made up mainly of
scents. I can find no evidence that they hear anything at all. It is true they can see near objects,
and are expert at distinguishing flowers by their
colours; but darkness prevails inside the nest, and
here everything is perceived, so far as one can tell,
by the senses of smell and touch, both of which are
conveyed through the antennæ, these organs being
in constant motion, investigating any object to
which attention is being paid, whether it be honey,
pollen, brood, comrade, or nest material. It cannot
be doubted that these and many other things that
have little or no smell to us are recognised by their
different odours. Humble-bees can readily distinguish the smell of their own species and that of
other species with which their lives are connected
in places that have been frequented by them, and
B. terrestris is almost as quick as the honey-bee to
discover honey or syrup. They resent, with angry
buzzing, the least whiff of human breath in their
nest, so the observer should breathe through a
corner of his mouth, a habit easily acquired. Yet,

in the case of most species, if the nest is uncovered cautiously, the moving of the material and the sudden flood of light appear not to be noticed, though the bees gradually realise that something is wrong, and, if left long exposed, will run off in search of material with which to re-cover the nest.

Humble-bees possess, in common with other

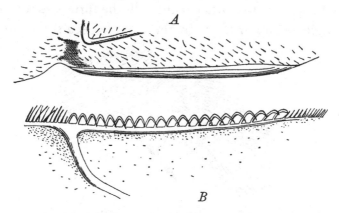

FIG. 1.—*A*, Portion of posterior margin of fore-wing of *Bombus muscorum* queen, seen from beneath. *B*, Corresponding anterior margin of hind-wing, seen from above.

bees, various peculiarities of structure adapted for special uses that it is hardly within the scope of this monograph to describe in detail, but a few of the most interesting may be here referred to.

First may be mentioned the wing-hooks, and I cannot do better than quote Bingham's description of these : " The winged hymenoptera are, as a rule, capable of swift and sustained flight. For this purpose they possess a wonderful arrangement (one of the most beautiful in nature) for linking together, during flight, the fore- and hind-wings. Examined with a good lens, the fore-wing is seen to have a fold along its posterior margin, while on the anterior margin of the hind-wing a row of hook-shaped bristles or hairs can be easily detected. When the wings are expanded these hooks catch on firmly to the fold in the fore-wing, and the fore- and hind-wing on each side are enabled to act in concert, having the appearance and all the firmness of a single membrane."[1] In the humble-bees the wing-

FIG. 2.—Antenna-cleaner in fore-leg.

hooks number from 17 to 24. Their position is indicated in Fig. 25, p. 145.

Another remarkable structure is an antenna-cleaner in the front legs. This consists of a semicircular incision in the metatarsus, fringed with a fine comb. When the leg is flexed a knife-like spine hinging from the tibia can be made to shut down over the incision. Thus a hole is formed through which the antenna is frequently drawn to rid it of any pollen grains or particles of dust that may have clung to it.

[1] *Fauna of British India—Hymenoptera*, by C. T. Bingham, 1897, vol. i. pp. viii and ix.

The tongue (Fig. 25, p. 145), using the term in its wider sense, is a complex structure consisting of an outer and an inner pair of sheaths, the maxillae and the labial palpi, which enclose the true tongue, this being in the humble-bee and its allies a long hairy organ having a groove on its under side. The honey is sucked up by the dilatation and contraction of the groove and of the tube made by the sheaths around the tongue. When not in use the whole apparatus is neatly folded away under the head.

Humble-bees breathe, not as we do through openings in the head, but through small holes in the sides of the body, called spiracles, of which there are two pairs in the thorax and five pairs (in the male six) in the abdomen. The spiracles of the thorax, which are situated under the wings, contain a vocal apparatus which is the source of the buzzing sound made by the humble-bee when it is irritated. Just inside the spiracle the windpipe is enlarged to form a sounding-box, and the sound is produced by the air expired passing over the edge of a curtain-like membrane fixed across the mouth of the sound-ing box. During the buzzing the wings, it is true, vibrate or quiver and increase the sound, but if they are removed the sound is still produced, while if the thoracic spiracles are covered, as Burmeister showed, the buzzing ceases or becomes so feeble that it is scarcely perceptible.

Other organs will be considered as occasion arises.

II

LIFE-HISTORY OF *BOMBUS*

THE story of the life of the humble-bee is largely that of the queen. From start to finish she is the central and dominating personage upon whose genius and energy the existence of the race depends. For she alone survives the winter, and, unaided, founds the colony in which she takes the position of its most important member.

The queen is raised in company with many others in July or August, the rearing of the queens being the final effort of the parent colony.

Fertilisation is the first important event in the queen's life. This takes place in the open air as a rule, but there are good reasons for supposing that it can be accomplished within the nest. The young queens are shy and show themselves very little. The males course up and down hedgerows, or hover over the surface of fields and around trees, in the hope of finding their mates. Each species has its particular kind of hunting-ground.

If, on a warm day in the month of July, one takes up a position, out of the wind, in a partial clearing in a wood, near some bushes or trees, one is almost sure to see now and then a humble-bee fly swiftly by and enter a dark hollow under a tree or shrub, where it pauses for a second, almost alighting, and then passes out and proceeds to another recess, where it again pauses and almost alights. Each succeeding bee flies in the same direction and visits the same spots. If these bees be caught, it will be found that they are all males either of *B. pratorum* or *B. hortorum*. This strange behaviour of the male humble-bee has puzzled many observers, but I have noticed certain facts about it that point to an explanation. A sweet fragrance, like the perfume of flowers, is perceptible about the pausing places. This same fragrance may be detected in the scent produced by a male if he be caught in the fingers, although it is now blended with an odour like that of sting-poison emitted in fear. Evidently, therefore, the males emit the perfume in their pausing places; and I think it extremely likely that in doing so they attract not only one another, but the queens. The males of the one species do not pause at spots frequented by those of the other species, and we may infer from this that each species emits a different scent.

The males of all the species are more or less fragrant when captured, those of *B. latreillellus* and *B. distinguendus* being especially so. The scent, I find, proceeds from the head, probably from the mouth. At the end of August 1910, my study was most pleasantly perfumed day after day by the males in a nest of *B. lapidarius* that was standing on a table there.

The males of *B. derhamellus* often disport themselves around the nest, waiting for the queens to come out; those of *B. latreillellus* will also do this; and I have seen a male of *B. ruderatus* ride away upon a queen as she was flying from the nest.[1]

Immediately after fertilisation the queen seeks a bed in which to take her long winter sleep. The queens of some of the species hibernate under the ground, others creep into moss, thatch, or heaps of rubbish. I have found *B. lapidarius* and *B. terrestris* and occasionally *B. ruderatus* and *B. latreillellus* in the ground, *B. lucorum* and *B. hortorum* in moss, and *B. pratorum* sometimes in the ground, sometimes in moss.

My observations have been made chiefly on the underground-hibernating species, *lapidarius* and *terrestris*. Both species pass the winter in much the same situations, but *terrestris* likes best to burrow in ground under trees, while *lapidarius* prefers a more open position, almost invariably

---

[1] The Rev. A. E. Eaton observed the males of *B. mendax* in the Berner Oberland, at an altitude of over 6000 feet, "resorting to favourite spots to bask (a stone or a spot of bare ground), hovering with a gradual fall like a feather, that ends almost imperceptibly in a dead stop, and standing with wings half spread, ready to dart off in an instant at the least alarm, or the sight of any insect flying past." E. Saunders, in quoting this in the *Entomologist's Monthly Magazine* (April 1909, p. 84), called attention to the enormous eyes of the male of *B. mendax*, and to the fact that the males of the sand-wasp *Astatus*, whose eyes are so large that they unite, have exactly the same habit.

choosing the upper part of a bank or slope facing
north or north-west, though generally near trees.

In such banks I have sometimes found great
numbers of queens, chiefly *lapidarius*, and it is
easy to discover them, because in burrowing into
the ground each queen throws up a little heap of
fine earth, which remains to mark the spot until the
rains of autumn wash it away. The burrows are
only one to three inches long, and if the bank is
steep they run almost horizontally. They are filled
with the loose earth that the queen has excavated.
The queen occupies a spherical cavity having a
diameter of about $1\frac{1}{8}$ inch.

It is evidently damp and not cold that the queens
try to avoid. Indeed, the northern aspect shows
that they prefer a cold situation, and the reason is
easily guessed. The sun never shines on northern
banks with sufficient strength to warm the ground,
so that the queens do not run the risk of being
awakened on a sunny day too early in spring, for
the queen humble-bee is very susceptible to a rise in
temperature in the spring, although heat in autumn,
even should it amount to 80° F., will not rouse her
when once she has become torpid. The queen
easily takes fright while she is excavating her
burrow, and I find that many burrows are begun
and not finished.

The queen always fills her honey-sac with honey
before she retires to her hibernacle. This store of
liquid food is no doubt essential for the preservation
of life, and is especially needed, one would think,

during September, when the ground is often very dry and warm.

Although the young queens may sometimes be seen flying in and out of the parent nest, I find that the majority of them leave it for good on their first day of flight; and as they are only occasionally observed gathering nectar from the flowers, I think that many, having filled themselves with honey before they leave the nest, become fertilised on the same day and immediately afterwards seek their winter quarters.

During the first few weeks the queen sleeps lightly, and if disturbed, for instance, by a visit from an earwig, she wakes up, creeps out of her burrow and flies away; but when the weather grows cold she folds her legs and bends as in death, sinking into deep torpor, from which she is not easily aroused.

The period of torpor lasts about nine months. Early species that commence sleeping in July, such as *B. pratorum*, are astir as soon as March and April, while later kinds wait until May and even June.

On sunny days in March the queens of *pratorum*, *terrestris*, and other hardy species may be seen busily rifling the peach-blossom, willow catkins, and purple dead-nettle, but in the afternoon as the sun descends and the air grows chilly they creep into hiding-places, where they relapse into semi-torpor, remaining in this condition until a favourable day again rouses them into activity.

The weather improving, the periods of animation become more frequent and last longer. Now each queen sets to work to search for a nest in which to establish her colony. The nest is usually one that has been made and afterwards vacated by a field-mouse, vole, or other small mammal, and consists of fine soft fragments of grass or moss, or it may be leaves, woven into a ball with a small cavity in the middle. Most of the species choose a nest that is under the ground, access to which is obtained by a tunnel varying in length from a few inches to a yard or more, but generally about two feet. The remaining species dwell in nests on the surface of the ground hidden in thick grass or under ivy; these are often called "carder-bees" because they collect material from around the nest and add it to the nest, combing it together with their mandibles and legs. But some of the underground-dwelling species occasionally occupy nests on or near the surface, often in strange situations, such as under boxes or in old birds' nests, rotten stumps, or out-houses, while some of the surface-dwelling species are sometimes found inhabiting nests under the ground, reached by a short tunnel.

In places where there is much moss or soft dead grass the carder-bee queen may sometimes construct the entire nest herself. It often happens that the mouths of the holes leading to the underground nests are overgrown with grass or ivy and half closed with debris, consequently they are not easily discovered, and the queens of the underground-

C

nesting species may be seen throughout the spring
hovering over the ground in woods and meadows
making a diligent search for them ; now and then
the queen alights in a promising-looking spot and
makes a closer examination of the ground on foot.

Having found a suitable nest, the queen becomes
rather excited and visits it frequently. Her first
flight from her new home is a momentous one, for
from it she has to learn how to find her way back
again to it. Having accustomed herself to the appearance of the entrance by crawling around it, she
ventures to take wing and poises herself for a moment
facing the entrance. Then she rises slowly, and,
taking careful notice of all the surroundings, de-
scribes a series of circles, each one larger and swifter
than the last. So doing she disappears, but soon
she returns and without much difficulty rediscovers
the entrance. Similar but less elaborate evolutions
are made at the second and third departures from
the nest, and soon her lesson has been learnt so well
that her coming and going are straight and swift.

She now spends a good deal of time in the nest,
the heat of her body gradually making its interior
perfectly dry. If the nest has been long unoccupied
and is in bad repair, she busily sets to work to
reconstruct it by gathering all the finest and softest
material she can find into a heap, seizing and pulling
the bits of material with her jaws and passing them
under her body backwards with her middle and hind
pairs of legs ; then she creeps into the middle of the
heap and makes there a very snug and warm cavity,

measuring about an inch from side to side but only about five-eighths of an inch from top to bottom, with an entrance at the side just large enough for her to pass in and out.

In the centre of the floor of this cavity she forms a little lump of pollen-paste, consisting of pellets made of pollen moistened with honey that she has collected on the shanks (tibiæ) of her hind legs. These she moulds with her jaws into a compact

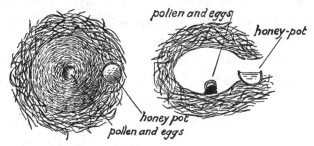

FIG. 3.—Diagram of commencing Nest.

mass, fastening it to the floor.   Upon the top of this lump of pollen she builds with her jaws a circular wall of wax, and in the little cell so formed she lays her first batch of eggs, sealing it over with wax by closing in the top of the wall with her jaws as soon as the eggs have been laid.   The whole structure is about the size of a pea.[1]

The method of collecting the pollen employed by the humble-bee and honey-bee, and the apparatus on the legs for carrying it out, are very wonderful and interesting; and as an essential part of the

[1] I think it likely that the eggs are sometimes laid in two lots, separated by an interval of a day or two.   Their number varies from 8 to 16; generally it is about 12.

process is not mentioned in our text-books on bees I will here refer to it at length.

Everybody has seen the loads of pollen, sometimes called wax in ignorance, on the legs of the bees. The load is carried on the outer side of the tibia or shank, which is concave, smooth, and bare, and fringed around the edge with long stiff hairs which act, as Cheshire observed, like the sloping stakes that the farmer places round the sides of his waggon when he desires to carry hay. This outer side of the tibia with its surrounding wall of hair is called the corbicula or pollen-basket.

In some flowers, such as wallflower and red ribes, the pollen is gathered by the mandibles, as noticed by Crawshaw, but in others it collects among the hairs of the body, especially those clothing the thorax and underside of the body, these being branched and thus admirably adapted for retaining it.

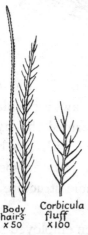

Body hairs × 50   Corbicula fluff × 100

FIG. 4.
Hairs of Humble-bee, magnified.

According to Hoffer (*Die Hummeln Steiermarks*, p. 37), the humble-bee brushes the pollen with the two first pairs of feet out of the body hairs forwards to the mouth, there chews and kneads it with honey and its saliva into a sticky paste, lays hold of it again with the feet, and presses it with the help of the middle legs on to the corbicula. But I believe the process is different in an important detail.

In the hind-legs, the next joint below the tibia, called the metatarsus, is enlarged into a sub-rectangular plate, which is densely clothed on the

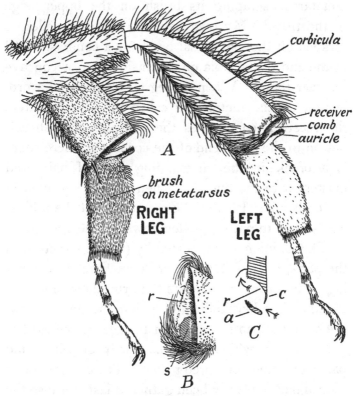

Fig. 5.—Pollen-collecting apparatus in Queen and Worker Humble-bee.

A. Hind legs of *Bombus terrestris* queen.

B. End view of apex of tibia, showing arch of hairs covering entrance to corbicula. *r*, receiver ; *s*, juncture of metatarsus with tibia (this is a ball-and-socket joint, the socket being here shown).

C. Diagrammatic section of receiver and auricle, showing method of working. *c*, comb ; *r*, receiver ; *a*, auricle.

inner side with stiff bristles forming a brush. In a pollen-collecting honey-bee this brush is filled with moistened pollen, which is evidently on its way to the corbicula. Cheshire states that the metatarsal

brush of the right leg transfers its pollen to the
corbicula of the left leg, and *vice versa*, but he
goes on to say that the transfer is effected by the
metatarsus scraping its brush on the upper edge
of the tibia.[1] My own belief is that the pollen is
scraped off the metatarsal brush by a comb, situated
at the end of the tibia on the inside, into a concave
receiver there. When the leg is straightened a pro-
jection on the metatarsus called the auricle enters
the receiver, compresses the pollen, and pushes it
out on to the lower end of the corbicula, where there
is a break in the surrounding wall of hair, and
plasters it to the mass of pollen already collected in
the corbicula. Finally, the metatarsus of the middle
leg is used to pat the pollen down on the corbicula.

This opinion is supported by (1) the structure of
the parts, (2) the fact that when the bees are collecting pollen from the flowers they rub their hind-legs
together in a longitudinal direction and do not cross
them, and (3) an examination I made of the load of
a honey-bee, which consisted partly of white and
partly of orange-coloured pollen. The orange pollen
(which had evidently been gathered last, because the
metatarsal bushes were filled with orange pollen)
was found only on that part of the corbicula that
was nearest to the auricle, where it had been forced
in as a wedge between the white pollen and the
corbicula, causing the whole mass of pollen to swell
and rise and also to buckle in the middle. The
outer side of the lump of pollen was tinged on the

---

[1] *Bees and Bee-keeping*, by F. R. Cheshire, 1888, vol. i. page 131.

surface with orange, showing where the metatarsus of the middle leg, which bore orange pollen grains, had patted it.

The same organs are present in the hind-leg of the humble-bee (see Fig. 5).[1]

To test the working of the apparatus I placed some pollen in the receiver of a leg of a dead queen of *B. lapidarius*, and then straightened the leg; the pollen was at once transferred to the corbicula.

When the pollen is being collected it always begins to gather at the lower end of the corbicula, and the reason is now clear. Also the smooth and slippery surface of the corbicula is explained, for the pollen slides up it, as the result of the numerous little contributions delivered on to it by the auricle.

The few hairs that obstruct the entrance to the corbicula number about three; they stand a little distance inside the entrance and are widely separated from one another. They provide a means of attachment for the pollen until the accumulated mass has grown large enough to be supported by the hairs at the sides of the corbicula. The edge of the entrance to the corbicula is densely clothed with fluff (seen under the microscope to be moss-like hairs), which probably serves the same purpose.

The surface of the receiver is smooth except

---

[1] In the honey-bee the bristles in the brush of the hind metatarsi are arranged in ten transverse rows, and are about as wide apart as the tips of the teeth of the tibial comb, but in the humble-bee they are not arranged in rows. In many pollen-collecting humble-bees the brush contains only a little dry pollen, and the moisture is confined to its upper corner, which is probably the only part that is much combed by the tibia.

along the margin bordering on the corbicula, where it is finely striate, the little furrows and ridges running in the direction in which the pollen moves. The auricle bears a tuft of hairs which helps to guide the pollen on to the corbicula.

Long hairs spring from either side of the entrance to the corbicula and form over it an arch which helps to support the accumulated mass of pollen without interfering with the delivery of fresh pollen from below. The arch is also of service in guiding the pollen on to the corbicula.[1]

Humble-bees working on the white dead nettle may be seen brushing the pollen out of the hairs on the front of the thorax, where it chiefly gathers, with the middle pair of feet, the instrument used being the metatarsus or basal joint of the foot, which is modified into a brush like the metatarsus of the hind leg. I have occasionally found a minute ball of moistened pollen in the mandibles, which seems to support Hoffer's view that the pollen is moistened in the mouth.

The wax of the humble-bee is much softer and

---

[1] See my paper, "How pollen is collected by the Social Bees, and the part played by the Auricle in the process," in the *British Bee Journal* for Dec. 14, 1911, and "Further Notes on how the Corbicula is loaded with Pollen," in the *B.B.J.* for April 11, 1912. In the latter article the receiver is named the *excipula* (*Lat.* receptacle), and the entrance to the corbicula the *limen* (*Lat.* threshold). In *Bombus confusus*, a native of Central Europe, the obstructing hairs on the limen are reduced to one ; in the honey-bee the fluff on the limen is scanty, and there are no obstructing hairs, except one, situated some way inside the entrance. In the humble-bee the working surface of the auricle is finely rugose, but in the honey-bee it is covered with pointed teeth inclining in the direction that the pollen moves. In *Bombus lapidarius, terrestris, lucorum, pratorum, lapponicus, latreillellus,* and *distinguendus* the auricle is hairy on the inner side ; in all the other British species it is bare there.

more plastic than beeswax, and is of a brown colour.
Hoffer considered that it was produced, like the
wax of the honey-bee, from the under side of the
abdomen, but I find that it exudes from between
the segments on the upper side of the abdomen ;
from here it is collected on the brushes of the hind
metatarsi (see pages 260-2). How it is conveyed
to the mandibles I have never been able to trace,
but that it is often dropped on to the comb and
afterwards picked up I have proved by placing a
grating of wire-cloth under a nest of *lapidarius* ; in
three days about 150 minute particles of wax had
fallen through the grating.

The eggs of the humble-bee are white and trans-
lucent, and shaped like a sausage, but slightly thicker
at one end than at the other. They are much
larger than the eggs of the honey-bee, their length
being $2\frac{1}{2}$ to 4 millimetres (about $\frac{1}{8}$ inch). Those of
the large underground-dwelling species are the
longest, being about three times as long as they are
broad ; those of the carder-bees are not only shorter
but stouter, their length not exceeding $2\frac{1}{2}$ times
their width.

The queen now sits on her eggs day and night to
keep them warm, only leaving them to collect food
when necessary. In order to maintain animation
and heat through the night and in bad weather when
food cannot be obtained, it is necessary for her to
lay in a store of honey. She therefore sets to work
to construct a large waxen pot to hold the honey.
This pot is built in the entrance passage of the nest,

just before it opens into the cavity containing the
lump of pollen and eggs, and is consequently
detached from it.

The completed honey-pot is large and approximately globular, and is capable of holding nearly a
thimbleful of honey.   Several honey-pots from nests
of *B. lapidarius* that I measured were ⅝ in. to ¾ in.
high and ½ in. to ⅝ in. in diameter at the greatest

FIG. 6.—Honey-pot from Nest of   FIG. 7.—The same, side view.
    Bombus lapidarius, seen from
    above.   Natural size.

width, but I believe these dimensions are sometimes
exceeded.

The honey-pot is exceedingly delicate and fragile,
the wall being thin as well as soft; but being left
undisturbed, it remains water-tight for about a
month, which is as long as it is needed.

The queen re-shapes the mouth of the honey-
pot daily according to her requirements.   Thus, at
night, when the honey-pot is full of honey, the mouth
is attenuated and small—only about a quarter of an

inch across, but in the morning after it has been
emptied the orifice is larger, the height of the pot
being reduced. The honey contained in the honey-
pot is of much thinner consistency than that stored
by honey-bees, from which a great deal of water has
been removed by evaporation before it is sealed
over. On a favourable day the queen fills the
honey-pot in a very short time; in my nests of
B. lapidarius I have often found it brimful at
9 o'clock in the morning. It takes the queen several
days to complete the building of the honey-pot
because of the large amount of wax it requires, but
the honey is deposited in it as soon as it is large
enough to hold any. Before the honey-pot is made,
or afterwards, if the honey-pot is full, the queen may
occasionally discharge her load of honey on to the
floor, and sometimes also on to the roof of the nest.
Although the honey soaks into the nest-material she
is able to suck up much of it during the night; what
remains on the floor is evaporated by the heat of
her body, leaving a sticky residue which glues the
nest-material together and helps to make it impervious to moisture.

In several nests of *B. lapidarius* and one of *B.
hortorum* I found that the construction of the honey-
pot was not begun until after the lump of pollen had
been made, and I think that this is usually the case.

The eggs hatch four days after they are laid.
The larvæ are maggot-like, being hairless and leg-
less, and as they begin to grow they assume a curled
shape, which is maintained until they are about to

pupate. When very small they are yellowish white, but this colour soon changes to white with a grey-pink tinge imparted by the internal vessels and their contents.

The larvæ, however, like the eggs, cannot be seen, for the queen keeps them covered with wax, in which they are completely enclosed as in a bag or skin.

The larvæ devour the pollen which forms their

FIG. 8.—Eggs, Larvæ, and Pupæ of *Bombus terrestris*, slightly enlarged.

bed, and also fresh pollen which is added and plastered on to the lump by the queen. The queen also feeds them with a liquid mixture of honey and pollen, which she prepares by swallowing some honey and then returning it to her mouth to be mixed with pollen, which she nibbles from the lump and chews in her mandibles, the mixture being swallowed and churned in the honey-sac. To feed the larvæ the queen makes a small hole with her mandibles in the skin of wax that covers them, and injects through her mouth a little of the mixture amongst the larvæ, which devour it greedily. Her abdomen contracts suddenly as she injects the food, and as soon as she has given it she rapidly closes up the hole with her mandibles. While the larvæ remain small they are fed collectively, but when they grow large each one receives a separate injection.

As the larvæ grow the queen adds wax to their covering, so that they remain hidden. When they are about five days old the lump containing them, which has hitherto been expanding slowly, begins to enlarge rapidly, and swellings, indicating the position of each larva, begin to appear in it. Two days later, that is, on the eleventh day after the eggs were laid, the larvæ are full-grown, and each one then spins around itself an oval cocoon, which is thin and papery but very tough. The queen now clears away most of the brown wax covering, revealing the cocoons, which are pale yellow.

These first cocoons number from seven to sixteen, according to the species and the prolificness of the queen. They are not piled one on another, but stand upright side by side, and they adhere to one another closely, so that they seem welded into a compact mass. They do not, however, form a flat-topped cluster, but the cocoons at the sides are higher than those in the middle, so that a groove is formed; this groove is curved downwards at its ends, and in it the queen sits, pressing her body close to the cocoons, and stretching her abdomen to almost double its usual length so that it will

cover as many cocoons as possible; at the same
time her outstretched legs clasp the raised cocoons
at the sides. In this attitude she now spends most
of her time, sometimes remaining for half-an-hour
or more almost motionless save for the rhythmic

FIG. 9.—Nest of *B. terrestris* showing the groove in which the Queen sits.
In the Frontispiece the Queen is seen incubating the brood.

expansion and contraction of her enormously distended abdomen, for nothing is now needed but
continual warmth to bring out her first brood of
workers.

In every nest that I have examined the direction
of the groove is from the entrance or honey-pot to
the back of the nest, never from side to side. By

means of this arrangement the queen, sitting in her groove facing the honey-pot—this seems to be her favourite position, though sometimes she reverses it—is able to sip her honey without turning her body, and at the same time she is in an excellent position for guarding the entrance from intruders.

After a period of rest the larvæ change to pupæ, heads uppermost. About the twenty-second or twenty-third day after the eggs are laid the perfect worker bees are formed, and, biting a hole through the tops of their cocoons, they creep out, those in the cocoons in the middle of the groove, which have been kept warmest, emerging a day or two earlier than those at the sides. In the work of biting open her cell the emerging bee is generally assisted by the queen or workers, and she makes several attempts to get out before the orifice is large enough to permit her thorax to pass through.

Should the weather be cold or incubation be interrupted, the duration of all the stages of development is lengthened. Insufficient feeding also delays the larval stage. Thus the time occupied from the laying of the egg to the emergence of the bee sometimes extends to a month.

The coat of the freshly emerged bee is matted and stuck down with moisture, and is of a uniformly dull silvery-grey colour. Her legs are weak and unsteady, and almost the first thing she does is to totter to the honey-pot, where she slowly unfolds her proboscis and takes a sip of the life-supporting drink. Then, refreshed and strengthened, she returns to the brood and nestles under the warm body of her parent. About forty-eight hours only are needed for her to acquire the handsome, well-groomed appearance and the bright rich colours of her mother, whom, indeed, she now resembles in every way except in her diminutive size.

The worker humble-bee commences to fly and to collect honey and pollen at a very early age. On June 18, 1910, at 3 P.M., I happened to examine one of my *lapidarius* nests and saw that no workers had emerged. On June 20, at 8 P.M., four workers were found in the nest, and one of them had already got her full colour. On June 21, at 10.30 A.M., a worker, evidently this one, was seen entering the nest with pollen on her legs, so that she must have been working in the fields when less than three days old. The honey-bee worker does not, as a rule, begin to gather honey until about the four-teenth day after she has emerged. The habit of collecting honey from the flowers comes by instinct, and is not the result either of experience or learning.

When the larvæ are spinning their cocoons the queen lays some more eggs, placing them, as always, in a little waxen cell, which, however, is now con-structed in a convenient place on top of one or two of the cocoons that form the sides of the depression in which she sits. Further batches of eggs are laid at intervals of two or three days, so that in a short time both sides of the groove are covered with cells containing eggs and young larvæ. All the eggs are laid on one side of the groove, usually two or three

cellfuls, before any are laid on the other side. It is
an interesting fact that the cocoons on which these
eggs are laid are not absolutely upright but are inclined inwards (see diagram), and the egg-cells are
constructed, not on the heads of the cocoons, where
they would hinder the exit of the emerging bees,
but on their outer sides.

As each cell seldom contains less than six, and
sometimes twelve or more eggs, a large family is
soon in course of development.

FIG. 10.—Diagram of the initial stages of the Humble-bee's Brood,
shown in vertical section.

The larvæ that hatch from the first of these
batches of eggs are generally approaching the most
rapid period of their growth when the first workers
emerge, and so the services of the latter commence
at the very time they begin to be most needed.

A little reflection will make it clear that the
cocoons in the first cluster and the cells of eggs
upon them are arranged in the best manner that
could be devised for deriving the most heat from
the queen's body. The cocoons are pressed close
together, those in the middle often assuming a
hexagonal shape (seen by cutting the cluster horizontally into two parts), like the cells of a honey-bee's

comb. The interstices between the tops of the cocoons are filled with wax, the surface of which is beautifully polished, making the groove in which the queen sits smooth and comfortable.

Four or five weeks of labour have told heavily upon the queen; the tips of her wings have become torn and tattered, and when she goes out to gather food she works less energetically than formerly, often stopping to rest on the leaf of a tree or on a blade of grass. As soon as she finds that her children are able to collect sufficient honey and pollen for the maintenance of the little family, she relinquishes this labour, and henceforth devotes herself entirely to indoor duties, laying eggs in increasing numbers and assisting the workers to incubate and feed the brood. Sometimes, however, the workers of the first batch are not sufficiently large or numerous to support the colony. In this case the queen continues going out to work until more workers appear.

A queen of a prolific species like *lapidarius* or *terrestris* during her most productive period lays a batch of eggs, on an average, daily.

The queen builds the special cell of wax to receive her eggs upon a cluster of cocoons, generally forming it in a crevice where two or three cocoons meet. I have, however, known a very prolific *lapidarius* queen to construct an egg-cell on a wax-covered cluster of nearly full-grown larvæ.

It is very interesting to watch a *lapidarius* queen lay a batch of eggs. An hour or two previously she commences to make the cup to receive them by collecting wax with her jaws, and depositing it in the selected place in the form of a ring, enclosing a space $\frac{3}{16}$ in. to $\frac{1}{4}$ in. across, the size depending on the number of eggs to be laid. On this foundation she builds the wall to a height of about $\frac{3}{16}$ in., no wax being placed in the bottom of the cup. At first the building is carried on intermittently and in a desultory manner, but an increasing amount of attention is paid to it, and the final touches are applied hurriedly. She gives the observer the impression that she is afraid the cell will not be finished before she has to commence laying. Suddenly she leaves off work, and, turning round, places the tip of her abdomen into the cup, which she clasps with her hind feet. In this position she remains for three or four minutes, her sting appearing through the wall of the cell every time an egg is laid. Directly she has finished laying, she turns round again, and, with her jaws, busily closes in the edge of the cup, and so seals it with a round and smooth covering of wax, a proceeding that occupies only a few seconds.

Each batch of eggs swells into a wax-covered bunch of larvæ, and finally becomes a cluster of cocoons. These clusters of cocoons lack the peculiar shape of the first cluster; not only is there no groove across the middle, but the cocoons in the centre are considerably higher than those at the sides, also the cocoons are less closely huddled together and the wax is more completely cleared away from them. The cocoons, too, are slightly larger, for the larvæ, nourished by so many nurses, grow to a larger size and develop into larger and stronger workers.

As the cell swells with the growing larvæ the bees are careful not to permit its area of attachment to the cocoons to grow larger, and they keep clearing the wax away from here, while to prevent it from falling over they fasten it by two or three pillars or ties of wax to adjacent cocoons or to the roof of the nest.

The larvæ of *B. terrestris* and *lucorum* do not keep together in a compact mass, but as they begin to grow large each one acquires its own covering of wax, although they do not separate completely ; the cocoons, therefore, do not form definite clusters, and are easily detached from one another. On the other hand, the bunches of larvæ and clusters of cocoons of *B. sylvarum*, *agrorum*, and *helferanus* are very compact and globular, and are often arranged in a ring around the centre of the nest, crowning the already vacated cocoons, and giving the comb a beautifully symmetrical appearance. The larvæ and cocoons of *lapidarius* also form compact masses, but two or more batches of worker brood, of nearly the same age, often coalesce as the result of their egg-cells being placed in line in contact with one another, consequently the clusters are often large and irregular.

With most of the species the skin of wax that covers each batch of larvæ is to the unaided eye unbroken, but as the larvæ grow, *B. terrestris*, *lucorum*, and *latreillellus* leave visible holes in the wax, which, when the larvæ approach full size, become large. The larvæ would now run the risk of falling out of their soft wax covering, which would mean their destruction, for a naked larva is always carried out of the nest; but they avoid this danger by enclosing themselves in a loose web of silk, doing this a day or two before they begin to spin their cocoons.

A larva that happens to lie underneath a large number of others generally has to build its cocoon almost horizontally, so that the end through which the perfect bee will escape may be free. Such a larva often fails to obtain a sufficient supply of food, with the result that it does not grow to full size and develops into a small bee. In this way tiny workers—I have seen some no larger than a house-fly—are sometimes produced, particularly in the nests of the carder-bees, who do not feed their young with such care as the underground species. If the worker is much undersized, her wings are almost sure to be malformed, so that she is never able to leave the nest.

The workers do not continue to use the queen's honey-pot, but, leaving it to grow mouldy and decay, they store the honey they gather in the cocoons they have vacated, adding wax to the rims of these to increase their capacity. They also, when the cells are full, reduce the size of their mouths with wax. The carder-bees, which produce little wax, usually rely entirely on these vacated cocoons for the storage of their honey, but some of the underground species, namely, *B. lapidarius, terrestris,* and *lucorum,* construct numerous waxen honey-pots as well. At first only three or four of these honey-pots are made, and they are a good deal narrower and less capacious than the queen's honey-pot, but as the colony grows

FIG. 11.—Comb of *B. lapidarius,* showing two honey-pots brimful of honey.

they are heightened, and their number is increased, and may amount in a large nest to twenty or even thirty : they are constructed at the side of the comb, and are usually joined together, forming a single or double row. In two colonies of *lapidarius* and one of *terrestris,* in which I was able to find the remains of the queen's honey-pot, I noticed that the first new honey-pots had been built on top of it. The honey in the honey-pots is always thin, showing

FIG. 12.—Comb of *B. lapidarius*, showing group of half-full honey-pots and irregular clusters of worker cocoons.

that it is freshly gathered and consumed every day, while the honey in the cocoons is thick, sometimes exceedingly so, showing that it is stored in these for use in times of scarcity. As the larger and newer cocoons become available for the storage of food, the oldest ones at the bottom of the comb are emptied and used no more, except in a time of plenty, when all the rest are full. In underground nests, where all available space is likely to be needed for the expansion of the comb, the walls of these abandoned cocoons are often bitten down, and the comb sinks. Wasps, it is well known, enlarge their nest cavity according to their requirements by digging out little lumps of earth and flying away with them, but I have never seen humble-bees do this.

The pollen, which is really a stiff paste of pollen and honey, is never put into the same cells as the honey. During the feeding of the first larvæ the queen deposits her pollen around the cell that contains them; here it is soon consumed, so that no receptacle is needed or made for it. Later on, however, as the comb grows, the pollen is placed in special cells, the nature of which depends upon the species. *Lapidarius* stores it under the brood in vacated cocoons, and sometimes also in small waxen cells. *Terrestris* and *lucorum* store it in one or two, afterwards increased to three or four, large waxen cells, which are built, sometimes singly, sometimes joined together, on top of the cocoons about, or not far from, the centre of the comb; these waxen cells, as the comb grows, rise like columns to a great height, towering above the brood, and contain an immense accumulation of pollen. But the carder-bees, and three or four underground species related to them, build little pouches or pockets of wax on to the sides of the wax-covered bunches of larvæ to receive the pollen, which, in

FIG. 13.—Photograph of a comb of *B. lucorum*, showing pollen-pots and honey-pots. The four pollen-pots rose ½ in. above the rest of the comb, and each of them was about 1¾ in. high, and ½ in. in diameter. The diameter of the comb was 4½ in.

their case, is gathered in comparatively small quantities. Upon this difference in instinct I have grounded a division of the species, calling those that store the pollen in cells detached from the bunches of larvæ "**pollen-storers,**" and those that place it in receptacles formed in the sides of the bunches of larvæ "**pocket-makers.**" [1]

[1] *Entomologist's Monthly Magazine* for the year 1899, p. 230.

This difference is really an important one, for the placing of pollen in contact with the brood is a vestige of the method of feeding employed by the solitary bees, which lay their eggs on lumps of pollen that they have collected, the larvæ feeding themselves on the pollen ; and I have no doubt that

FIG. 14.—Comb of *B. agrorum*, showing pollen-pockets in the sides of the bunches of larvæ.

the larvæ of the pocket-makers do partly feed on the pollen placed in the pockets, at least during the earlier stages of their growth. It is only the pollen-storers that, after the first batch has been reared, quite abandon supplying their young with solid pollen, and completely adopt the honey-bees' more advanced method of feeding with liquid food prepared in the body of the bee.

The underground pocket-makers, namely, *ruder-atus*, *hortorum*, and *latreillellus*, and probably also *distinguendus*, a close relation of *latreillellus*, prime their egg-cells with pollen, placing a little pollen in the bottom of each egg-cell constructed, and laying their eggs upon it, and thus they preserve another trace of the primitive feeding-method of the solitary bees. These four species form a natural group, which I propose to call "**pollen-primers.**" They are all large, with long heads and long tongues.

The pocket-makers, at least *B. ruderatus*, *der-hamellus*, *agrorum*, and *helferanus*, make, as a rule, only one pollen-pocket for each bunch of larvæ, but, under certain conditions, they may make two or three. When many larvæ are being reared by a small staff of workers, the pockets are small, and the pollen that is placed in them is plastered on to the wall covering the larvæ, from which it quickly disappears, being no doubt consumed almost im-mediately ; but when the population is greater, and the weather being fine, pollen is gathered in plenty, the pockets are large and cup-shaped, and contain during the day a shallow store of pollen, the surface of which is concave. In a nest of *derhamellus* that I took on July 2, 1911, in the height of prosperity, the pollen-pockets or cups were very large and of an oval shape, several of them measuring $\frac{5}{8}$ in. long by $\frac{1}{2}$ in. wide ; they were, however, quite shallow, the depth of the pollen in them at the centre being only about $\frac{1}{8}$ in., and there was only one pocket to each bunch of larvæ. When the larvæ are full grown the pollen-pockets are destroyed.

When the usual receptacles for pollen employed by a particular species are not available, it may adopt those employed by others. Thus in a strong nest of *B. agrorum*, one of the pocket-making species that I had under observation in 1910, the workers, during a period when there were no growing larvæ and consequently no pockets for pollen, dropped all the pollen they brought home into a special waxen cell they had constructed, like *terrestris*, on the top of some cocoons. Also a colony of *B. hortorum*, another pocket-maker, being in an advanced stage, and having no growing larvæ, placed pollen in the cocoons vacated by the young queens, but only lined the interior of the cocoons with it.

In general, humble-bees, like honey-bees, prefer to deposit the pollen in cells among the brood, and the honey in cells farther off.

It will be understood that the comb expands in an upward and lateral direction. At the bottom of the nest are the vacated cocoons, now filled with honey ; lightly resting on these, with narrow gang-ways for the bees in every direction between them, are the clusters of cocoons containing pupæ and full-fed larvæ. Amongst and above these are the bags of wax of various sizes containing larvæ in different stages of growth. Finally, here and there, on the clusters of cocoons containing the pupæ and larvæ are the little sealed waxen cells containing eggs. No brood is therefore visible.

The nest material is pushed out to make room for the growing comb, and there is always a space for the passage of the bees between it and the top of the comb.   Over this space, and lining the inside of the nest material, a ceiling of wax is generally made in populous nests, the wax being worked into a thin sheet as in the coverings of the larvæ and the walls of the honey-pots.   Provided there is room for it, this waxen canopy is always found completely enveloping the top of the comb in populous colonies of *B. lapidarius*, an underground-dwelling species that produces more wax than any other. Populous colonies of the underground species *terrestris*, *lucorum*, *ruderatus*, and *hortorum* generally succeed in time in making complete canopies, but in the nests of the surface-dwelling species, which, one would imagine, specially need such a protection to keep out the rain, the waxen ceiling is, in most cases, incomplete or absent, and frequently consists only of a small disc of wax over the centre of the comb. *B. derhamellus* is particularly disinclined to commence building a ceiling; in several very populous nests of this carder-bee no trace of one was seen.

In order to watch what is taking place in my observation nests I have sometimes had to remove the waxen ceiling.   In a populous colony of *lapidarius* it has been entirely built again within two days, the construction commencing at the sides of the box, and the whole ceiling built from these with no other support but that of the bees constantly passing to and fro on the comb.

Succeeding batches of workers emerge in due time and the population of the colony grows rapidly, a few fresh workers emerging every day : these workers are still larger than the earlier ones, and they are very capable and energetic.

Every active moment in the worker's life, which lasts about four weeks, is employed in furthering the prosperity of the colony. Even before her full colours have appeared she begins to nurse her baby sisters, spreading her body over them and feeding them. The adult worker spends the greater part of the day journeying to and from the flowers, and she seldom returns home without her abdomen distended with honey, and the tibia of each of her hind legs bearing a large pellet of pollen.

It is interesting to watch the worker put away her load. After entering the nest she runs about, feeling and smelling with her antennæ, in search of cells to receive it. Having discovered a receptacle containing pollen, she takes a step forward so as to bring her hind legs exactly over the mouth of it, and rubbing them together, she detaches the two pellets, which drop into the cell. Then she may turn round and, putting her head into the cell, may spread and plaster down the pollen with her mandibles, but often she leaves this to be done by another worker. Hastening to a cell containing honey she buries her head in it, and her abdomen is seen to contract as she regurgitates the honey. Next minute she is out and away to collect another load.

She does not always drop her pellets of pollen at

the first attempt, but, just as she is about to dislodge them she appears to be often seized with doubt as to whether they will fall into the cell, and, turning round, puts her head into the cell : having thus re-ascertained its exact position, and perhaps reassured herself of its suitability, she again steps forward, and this time lets fall her load.   The queen, should she be near, is much interested in the arrival of the pollen, and on one occasion I saw her nibble some of it off the worker's leg while the latter was engaged for a moment with her head in the cell.

When the weather is warm the workers are particularly industrious in the cool of the evening, bringing home heavy loads until dusk.   At night the colony is even more animated than in the day-time, for the whole population is now at home, and each bee is occupied, some building, some feeding the larvæ, but the great majority slowly creeping over the brood in all directions, stopping now and then for a moment or two to spread their bodies over some portion of it.   No special attention is paid to the queen.

The surface-dwelling species often pass in and out of the nest through a covered way under the grass, more or less lined with fragments of dead grass or moss, and extending from four to twelve inches from the nest.

The nests of the underground species are badly ventilated, and in populous colonies, during the heat of the day, one or more workers will station them-selves on top of the comb or at the lower end of the

tunnel, there maintaining a continual fanning and
humming with their wings.   Once, on a still day, I
discovered a nest of *B. pratorum* through the humming of a ventilating bee in the tunnel.   Around
the fact that the fanning and humming are often
started early in the morning by a single bee standing
upon an eminence of the comb, early observers
wove a pretty story that the humble-bees possess a
trumpeter or drummer who at a certain hour ascends
to a box or stand contrived for the purpose on the
summit of the comb and sounds a reveillé, calling
the inhabitants to begin the day's toil.[1]

[1] The following abridged extract from *Nature Display'd*, a translation
by Samuel Humphreys of *Le Spectacle de la Nature*, by the Abbé Noël Pluche,
published in several editions, 1732 to 1750, shows the quaint ideas of this
period about the behaviour of humble-bees in their nest :—

"I have seen amongst my wild Bees, and that very frequently, a large
Insect, much superior in Size to the rest ; it was as bare as a pluckt Fowl and
black as Jet or polished Ebony.   This King goes from time to time to survey
the Work ; he enters into each particular Cell, seems to take their Dimensions
and examine whether the whole be finished with due Symmetry and Proportion.
I am very apt to suspect this Monarch to be a Queen and that her Visits to
each Cell only tend to deposit her Eggs there.   When she makes her publick
Appearance all the young Bees who form her Court plant themselves in a
Circle round about her, clap their Wings, raise themselves on their fore Feet,
and after several Leaps and Curvets and other Expressions of their Joy, attend
her throughout her Progress, at the Conclusion of which the Queen retires and
all the rest return to their Employment.

"In the Morning the Young appear indolent, and are with great Difficulty
brought to apply themselves to their several Functions: in order to rouse them,
one of the most corpulent of their Band, exactly at half an Hour past Seven,
erects his Head and part of his Body out of a Box or Stand contrived for that
Purpose ; there he claps his Wings for the Space of a Quarter of an Hour and
with this Noise awakens all his People.   This summons them to work and is
the Drum to beat the Signal of their March.

"There is likewise another who keeps Guard all Day, and I have seen him
acquit himself of his Commission with a Vigilance that astonished me.   I have
sometimes thrown a common Bee into the Hive after I had plucked off one of
his Wings ; but he was instantly seized by the Centinel and laid dead on the
Spot.

"Four days ago, our Queen set out very early in the Morning, and
proceeded, infirm and old as she was, with a trembling March to the Confines of

Hoffer relates how his scepticism about the existence of the so-called trumpeter was suddenly dispelled by his discovering it at 3.30 A.M. in a very populous nest of *B. ruderatus*, variety *argillaceus*, he had set up in his window facing south-east. Morning after morning the trumpeter arose about this time and continued to hum for about an hour. He roused his wife and children, and they, too, saw and heard the trumpeter.

After the queen has laid altogether from 200 to 400 eggs that will develop into workers, the number depending upon the species and the vigour of the queen, she begins to lay others that are destined to produce males and queens.

It is almost impossible to distinguish the male brood from the worker brood, but the full-grown queen larvæ and the queen cocoons may be known by their larger size. Sometimes the cocoons in each cluster contain males only or queens only, but often the two sexes are mixed. As a rule, the earlier batches produce chiefly or only males, and the later ones chiefly or only queens. In some nests, however, all the eggs are female, not a single male being produced. This happened in a strong colony of *B. pratorum* that I kept under observation until it died out. In other cases males only may be produced,

her Dominions. I saw her Majesty drop down behind a little Eminence, and, after having languish'd a short Time, she expired. The whole City was inconsolable on the Occasion. The Drum did not beat the Signal that Morning as usual. A heavy Melancholy sat on every Brow. The number of the Inhabitants since that time has daily diminished, and they are continually removing in Quest of a new Settlement."

E

as I once noticed in a nest of *B. terrestris*. A few
males are often produced with the later broods of
workers, especially if the queen is not prolific, but
workers are seldom produced with the males and
queens, and such as do appear have probably failed
to develop into queens through insufficient feeding.
I have never known a queen to be produced among
the regular batches of workers, but in some of my
nests that I fed liberally large workers like those
that are produced in the later broods were produced
in the early broods. It seems, therefore, that an
abundant supply of food is not sufficient to make a
female larva develop into a queen, but that it may
also be necessary for the larva to be from an egg
that has been laid late in the queen's life. Also the
possibility that the development of a queen instead
of a worker is the result of a slight differentiation in
the food given to the larva must not be precluded.
With the honey-bee, if the female larva is fed entirely
on "royal jelly," a rich milky food prepared in the
chyle stomach of the bee, it develops into a queen ;
but if it is weaned on the third day, and thereafter
has honey added to its food, it becomes a worker.
With the humble-bee, although no such differentiation in feeding is observable, and the queen larva
appears to be fed like the worker larva on a gruel
of honey and pollen, prepared in the honey-sac, it is
not improbable that the composition of this food
is slightly altered, when the colony has reached a
certain stage, by, for instance, a greater activity in
some of the salivary glands, of which no less than

four systems are known to exist in the worker honey-bee.

The queen larvæ take longer to attain their full size than either the worker or the male larvæ. In a cluster of cocoons containing both males and queens the queens are on top and the males at the sides.

On an average, taking one nest with another, it may be estimated that nearly twice as many males as queens are produced. The total number of males and queens reared varies from 100 to 500, according to the staff of workers.

In a *lapidarius* nest a strange scene may be witnessed at the laying of the male and queen eggs. The workers, hitherto so amiable, are suddenly seized with anger and jealousy, for as soon as the queen has closed the cell and turned away, one or two of them hurriedly commence to bite it open, their wings quivering with excitement. The queen, however, seems to have expected this behaviour, and, quickly returning to the cell, throws down the conspirators, repairs the cell, and again departs. But directly her back is turned a worker again attacks the cell, and again the queen beats it off. And so continues a game of attack and defence for five or six hours, sometimes one worker, sometimes another, being the offender. At last the workers leave the cell unmolested. That their object is to destroy the eggs is proved by the fact that when sometimes a worker succeeds in reaching the eggs I have seen it seize one and devour it with much relish. In this way many eggs must be destroyed. I have also seen the workers of *B. terrestris* attack the new-laid male and queen eggs, but in a much less determined manner.

The young males leave the nest as soon as they are able to fly, and do not return again for food or shelter. Their life, though idle, is brief, and does not last more than three or four weeks. The young queens may sometimes be seen returning to the nest, occasionally with pollen on their legs, but they too soon leave it for good when they get mated, and seek their winter quarters.

As the old queen ages she gradually loses her hair. First, the greater part of the abdomen, and then the mesonotum, or central part of the thorax, become more or less bald. The queens of *terrestris* lose their hair more rapidly and completely than those of any other species that I have observed.

With advancing age the queen's prolificness falls off rapidly, and she is often scarcely able to lay enough eggs to keep the workers fully employed; some of the latter then lay eggs which, however, produce males only. A laying worker quickly loses much of the hair on its abdomen, and by this means may often be discovered. Some species, particularly *lapidarius* and *terrestris*, are more liable to develop laying workers than others, and in a normal nest of *terrestris*, even while the queen is still prolific, eggs may often be found in the undersized workers with malformed wings that never leave the nest; but I believe that these eggs are seldom developed and laid. Although I once caught a worker and male of *B. agrorum* in copulation, there is ample evidence to show that the laying workers are virgins, and this explains why they produce males only, the parthenogenetic production of the drone being a well-known phenomenon in the case of the honey-bee.

The eggs of the workers are as large as those of the queen ; even the tiny workers lay full-sized eggs, although they cannot develop more than one or two at a time. Several workers will lay their eggs in the same cell, and while they are ovipositing there is a great deal of rivalry and quarrelling between them. But unless the queen is unprolific, or dies early, the workers produce very few offspring, indeed in many nests they produce none.

The store of honey in the cocoons is generally at its greatest soon after the young queens begin to emerge. In a favourable season a populous colony may have all the vacated cocoons, amounting to over 400, filled with thick honey and sealed over with wax.

The largest colonies are made by *lapidarius*, *terrestris*, and *lucorum*. On July 24, 1894, I took a nest of *lapidarius* containing 245 workers. On July 14, 1911, I took a *terrestris* nest with 221 workers. In both these nests workers were still emerging from their cocoons, and the oldest workers had died. I have also taken very strong nests of *lucorum*, and I think that the number of workers in a nest of any of these species must sometimes reach 300. The colonies of the carder-bees, on the other

hand, are not nearly so populous, but I have taken
nests of each of the species containing over 100
workers. Many colonies, however, fail to grow
large through parasites, bad weather, and various
other misfortunes ; thus in some nests of the prolific
underground species the population never exceeds
100, and in some nests of the carder-bees it never
exceeds 50.

When most of the young queens have emerged
the number of workers diminishes rapidly, and is
soon reduced to a few dozen of the youngest and
strongest. These drop off one by one. Flowers
grow scarce, the workers become idle and listless,
and the store of honey that only two or three weeks
previously filled so many cells is quickly consumed.
The comb grows mouldy, and the old queen dies.
And so decay and death overtake the once busy
community.

In the case of *B. pratorum*, and probably of the
other species whose colonies end their existence in
the height of summer, the aged queen often spends
the evening of her life very pleasantly with her little
band of worn-out workers. They sit together on
two or three cells on the top of the ruined edifice,
and make no attempt to rear any more brood. The
exhausting work of bearing done, the queen's body
shrinks to its original size, and she becomes quite
active and youthful-looking again. This well-earned
rest lasts for about a week, and death, when at last
it comes, brings with it no discomfort. One night,
a little cooler than usual, finding her food supply

exhausted, the queen grows torpid, as she has done many a time before in the early part of her career; but on this occasion, her life-work finished, there is no awakening.

. . . . . .

Thus the life-history of the queen humble-bee is completed. It only remains to describe a slight, but nevertheless very interesting modification of it that sometimes takes place.

Many of the later appearing queens of *B. lapidarius* and *terrestris*, two of the most abundant species in England, do not take the trouble to start nests of their own, but finding a nest already occupied by a queen of their own species attach themselves to it. The foundress of the nest at first ignores the stranger, who takes care to keep out of her way as much as possible. After a short while, however, the intruder grows bolder, and begins to pay close attention to the brood. Jealousy then arises, and a mortal duel is the result. The two queens seize and endeavour to sting one another in the most ferocious and desperate manner, rolling over, locked in a deadly embrace. One of them succeeds, usually within a few seconds, in piercing the other, the sting in most cases penetrating between two of the segments of the abdomen. Instantly the sting of the wounded queen becomes paralysed, and she relaxes her grip of the victor. Then she grows cold and lethargic, and sometimes at once, sometimes an hour or two later, she dies. The foundress is generally the conqueror, as she deserves to be, but this is only because she fights with greater energy and boldness than her adversary, the combatant that shows fear, hesitation, or weakness being almost always defeated. Consequently, if the foundress has become enfeebled, or has been taken by surprise by the interloper, she is very likely to lose her life. The contest decided, the victor pays no further attention to her vanquished antagonist, and, with an air of relief and satisfaction, goes and cherishes the brood with the greatest affection.

In seasons when *lapidarius* or *terrestris* queens are abundant, several of these duels may be fought in a single nest, and in digging up nests in an early stage I have often found the dead bodies of the queens. They lie under the nest or near it, in the cavity or tunnel. I once dug up a nest of *terrestris* containing the remains of no less than twenty queens : the tunnel was short with a conspicuous entrance. On May 27, 1911, I found two dead *terrestris* queens in a nest with a short tunnel, the mouth of which was small and completely hidden under grass, which shows that the queens often find the nests by scent.

I think that the queens of the carder-bees seldom enter one another's nests and kill one another ; but that they may do so sometimes was proved by my finding a dead *derhamellus* queen, as well as the reigning one, in a *derhamellus* nest on June 3, 1911, this species being particularly plentiful that season.

It may be asked, What causes the queens to fight ? Investigations that I have made point to the conclusion that the original queen tolerates the intruder only so long as the latter has no eggs to lay, the two mothers being unable to endure one another's presence.

Each queen that attaches herself to the nest of another means one less colony started, and it is hard to see of what advantage this habit can be to the species except the survival of the best fighters. One would suppose that the attachment of one or more queens to a nest might be a safeguard against the perishing of the young, should their mother get lost, but my observations show that the satellite is not sufficiently settled in life to assume the care of the brood until she is about to lay eggs. It is, in fact, the imminence of motherhood that makes her look after the brood, although success in a duel rouses her to do so.

I have discovered one very interesting exception to the rule that only the queens of the same species prey upon one another.

On June 15, 1894, I found a nest containing a *terrestris* queen with ten workers, all of which were of the closely allied species *lucorum*, and a dead *lucorum* queen, evidently the mother of the workers. On the same day I found another nest containing about twenty workers, all *lucorum*, with a *terrestris* queen and the decaying remains of two *lucorum* queens. These, and several similar cases met with in later years, leave no doubt that the *terrestris* queens frequently enter the *lucorum* nests, kill the *lucorum* queens, and get the *lucorum* workers to rear their young. *Lucorum* is milder tempered and less vigilant than *terrestris*, and one can well imagine that in a contest between a *terrestris* and a *lucorum* the *terrestris* would be victorious.[1]

[1] See my article in the *Entomologist's Monthly Magazine* for 1899, p. 233.

III

PSITHYRUS, THE USURPER-BEE

THE most interesting and, in my experience, the deadliest enemies to which several of the commonest species of humble-bees are liable to fall a prey are bees so closely resembling the true humble-bees themselves that only a student can tell the difference between them.   These bees were first recognised to be distinct from the ordinary humble-bees in 1802 by Kirby, who noticed among other differences that the females lacked the pollen-collecting apparatus on the hind legs,[1] and thirty years later Lepeletier gave them the name of *Psithyrus*.

Each species of *Psithyrus* breeds only in the nests of its own particular species of *Bombus*.   The *Psithyri* produce no workers, and some of the early observers who saw them in the nests of the *Bombi*, noticing that they appeared to live in perfect harmony with their hosts, suggested that they might render them some important service.   But it was soon discovered that their association with the *Bombi* was to the disadvantage of the latter, and they came to be regarded as commensals living

[1] *Monographia Apum Angliæ*, vol. i. pp. 209, 210.

upon the bounty of the *Bombi*. This view was
upheld by Hoffer, who, in 1889, published a monograph on the *Psithyri*[1] in which he recorded observations of his own in Styria supporting it. But
my observations have shown that at least two of
the species of *Psithyrus* are deadly parasites upon
the species of *Bombus* in whose nests they breed.
In the years 1891 to 1894 I dug up many nests
of *Bombus lapidarius* containing its parasite
Psithyrus rupestris, and of *B. terrestris* containing
its parasite *Ps. vestalis*, and I noticed that in every
case the *Psithyrus* female had taken the place of
the mother of the colony, whose remains I generally found lying under or near the nest. Investigation showed that it is the practice of the *Psithyrus*
female to enter the nest of the *Bombus*, to sting
the queen to death, and then to get the poor workers
to rear her young instead of their own brothers
and sisters.[2]

The way in which the *Psithyrus* queen proceeds
in order to ensure the success of her atrocious work
has all the appearance of a cunning plan, cleverly
conceived and carried out by one who not only is
a mistress of the crime of murder, but also knows
how to commit it at the most advantageous time
for herself and her future children, compelling the

[1] *Die Schmarotzerhummeln Steiermarks*, by E. Hoffer (Graz. Nat. Ver.),
1889.

[2] An outline of the life-history of *Psithyrus* here described was given in my
privately published booklet *The Humble-bee, its Life-History and How to
Domesticate It*, in 1892. See also *Amateur Naturalist*, Croydon, of May
1893, p. 38 ; and *Hymenoptera Aculeata of the British Islands*, by E. Saunders,
1896, p. 354.

poor orphans she creates to become her willing slaves.

The following are the details of the life-history of *Psithyrus rupestris* and *Ps. vestalis* as I have observed them :—

The queens hibernate solitarily in the ground like the *Bombus* queens, but they do not quit their winter quarters until after the *Bombus* queens have emerged, and most of them are already engaged in rearing their first batch of workers.

Psithyrus rupestris much resembles her victim, the *lapidarius* queen, in appearance, being about the same size, and, like her, having a black coat with a red tail ; but her wings are dark brown, not clearly transparent as in *lapidarius*, and her flight is feeble, producing a lower note than that of the *lapidarius* queen, the rate of wing vibration being slower. *Ps. vestalis* also rather resembles her victim, *B. terrestris*, in her yellow band and whitish tail, but is likewise distinguishable by her smoky wings and soft low-sounding flight.

But the most remarkable feature about the *Psithyrus* queens is their exceedingly thick and hard skin, covering them like a coat of mail and protecting them from the stings of the *Bombi*. The segments of the abdomen, in particular, are very hard and lap tightly and closely over one another, there being no wax-yielding membrane between the dorsal segments, so that it is very difficult for an adversary to force her sting between them. Their coats are thin, perhaps as a compensation

for the extra thick skin, many thick-hided animals
lacking hair. Their stings are stouter and more
curved than those of *Bombus* : it is interesting to
note that the sting in the queen honey-bee, who
also uses it for killing queens, is curved, while in
the worker honey-bee the sting is straight.

The movements of the *Psithyrus*, whether flying
or walking, are lethargic and awkward. When
visiting the flowers in search of food she does not
travel systematically from blossom to blossom like
an industrious humble - bee, but settling upon a
bloom she sips lazily sufficient nectar to satisfy her
immediate need, and afterwards is very likely to
become drowsy. Fatigued by the exertion of
obtaining food for herself, she is plainly incapable
of the sustained effort that would be needed had
she to provide for the wants of her young.

Whenever the weather is pleasant she searches
leisurely for a nest of the particular species of
Bombus which it is her instinct to victimise. In
this work she is guided, like a dog, largely by
scent. In May 1895 I caught a searching queen
of *Ps. vestalis* and put her into a glass jar in which
I had kept some queens of its host, *B. terrestris*,
for several hours, so that the jar had acquired the
characteristic odour of this bee. The *Psithyrus*
queen ran about inside the jar in great excitement, waving her antennæ and stroking the interior of the jar with them, apparently trying to
trace the path the queens had taken. After a few
minutes' hunting she flew out of the jar, but finding

she had lost the scent she returned at once to search again inside.

The *Psithyrus* is much more likely to find a nest if the tunnel leading to it is short than if it is long. Most of the nests containing *Psithyri* that I have dug up had tunnels not exceeding fifteen inches in length, and in none were they over two feet. It is probably to escape the *Psithyrus* that *lapidarius* and *terrestris* nests often have longer tunnels than those of any other species.

In her wanderings the *Psithyrus* may find and enter the nest of some other species of *Bombus.* Having succeeded in making herself acceptable to the inhabitants, she becomes a temporary lodger in this nest, making it her headquarters and returning to it for meals and also to pass the night. I once found a *Ps. rupestris* queen lodging in a weak nest of *B. agrorum* in the thatch of a cow-lodge ; another *rupestris* was a frequent visitor in a nest of *B. derhamellus* I had under observation in June 1910, where she was supported by the queen only, no workers having yet emerged. I found a *Ps. vestalis* lodging with a *B. derhamellus* queen on June 9, 1910. On June 10, I found a *Ps. vestalis* in a nest of *B. pratorum* with the queen and seven workers ; and on July 4, I discovered a *Ps. vestalis* in a nest of *B. pratorum* containing the queen and twenty workers. I have never found a *Psithyrus* lodging in a populous nest.

The majority of the *Psithyrus* queens discover the nests of their victims when only the first batch of workers has emerged, and this is the most favourable time, for these early workers are less hostile to strange queens than are the workers that emerge later. If the *Psithyrus* fails to find the nest until the workers have become numerous they set upon her with great fury, and, after some time, generally succeed in killing her by stinging her in a vulnerable spot, as for instance in the neck. On June 20, 1894, I took a nest of *B. terrestris* in an unusually forward stage for this date, there being 50 workers, 6 males, 2 young queens, and a quantity of male and queen cocoons. In the hole were two dead *Psithyrus vestalis* females, both perfectly denuded of hair, and 15 dead *terrestris* workers. Evidently the *Psithyri* had been killed, but only after severe fighting and heavy loss. Again, on July 14, 1911, I found the dead body of a *Ps. vestalis*, also absolutely hairless, lying in the grass at the mouth of the tunnel of a strong *terrestris* nest : the workers must have killed her and dragged her to the surface.

It is, therefore, necessary for the *Psithyrus* to find the nest before many workers are out, but it would not be to her advantage to kill the queen until the latter has laid the greater number of the worker eggs ; and, as a matter of fact, she does not do so.

The *Bombus* queen, on first meeting the *Psithyrus* in her nest, shows a certain amount of agitation, and may advance to attack her, but, her courage failing, she draws back. The *Psithyrus*, however, treats the queen with good-natured indifference, unless the latter becomes disagreeably aggressive, and then all she does is to lift occasionally a warning leg, or to creep away and hide herself like a coward in the nest material. If the workers attack her, she tries to rub them off with her legs, and slips into a crevice between the clusters of cocoons, or into the nest material; but, protected by her coat of mail, she has little cause to fear getting stung.

Although the *Psithyrus* during the first few days flies occasionally to and from the nest, I have seen no evidence that she brings home any food, or that she helps in any way to rear the *Bombus* brood. Her first care is to ingratiate herself with the inhabitants, and in this she succeeds so well that the workers soon cease to show any hostility towards her. Even the queen grows accustomed to the presence of the stranger, and her alarm disappears, but it is succeeded by a kind of despondency. Her interest and pleasure in her brood seem less, and so depressed is she that one can fancy she has a presentiment of the fate that awaits her. It is by no means a cheerful family, and the gloom of impending disaster seems to hang over it.

But while the queen grows more dejected, the *Psithyrus* grows more lively, and takes an increasing interest in the comb, crawling about over it with unwonted alacrity, and examining it minutely.

In June 1908 I had a good opportunity of watching a *Ps. rupestris* establish herself in one of my nests of *B. lapidarius* that I had placed in

F

a box for observation. *Rupestris* queens were particularly plentiful that season, and one was first seen in the *lapidarius* nest when only eight workers had emerged. The next day when I looked into the box this *rupestris* was out, but on the following day I was surprised to see two *rupestris* on the comb, apparently living on good terms with the *lapidarius* queen and her children. After this I inspected the box at least twice every day. On the first few occasions there were sometimes two *rupestris* females, and once three, to be seen in the nest, but after two or three days only one remained, and she was always at home. The poor *lapidarius* queen was visibly depressed and ill at ease in the dominating presence of the *Psithyrus*. As time went on she grew nervous and languid, and showed increasing fear and suspicion of the unwelcome guest. Finally, at 11.30 A.M. on the tenth day after the first appearance of a *Psithyrus* in the nest, I found the queen and three workers lying dead outside the nest, and the *Psithyrus* on the comb displaying more activity and satisfaction than usual. Evidently the three workers had sacrificed their lives in a futile attempt to destroy their mother's murderess. The next day the *Psithyrus* laid some eggs.

Here we have evidence for supposing that the death of the queen is brought about by the development of eggs in the *Psithyrus* causing jealousy and a duel, the two queens, as in the case of rival *Bombus* queens, being unable to tolerate one another's presence when they have eggs to lay. Whether the jealousy is mutual, whether the *Psithyrus* is the aggressor, or whether the *Bombus* queen compasses her own destruction by first attacking the *Psithyrus*, it is impossible to say. In the above case, only three hours before I discovered that the death of the queen had occurred, I looked into the nest and saw both queens on the comb, and there was nothing in their behaviour towards one another to indicate the approaching tragedy. On the whole, I think it probable that the *Bombus* queen starts the fight as soon as she discovers that the *Psithyrus* is about to lay, an unpardonable fault in a member of her household at this early stage.

I believe that a *Psithyrus* queen never commences laying as soon as she enters the nest of her host, and that egg-production is the result of living in luxury in the nest for some days. Indeed, there is every reason to suppose that if the colony when discovered by the *Psithyrus* is in too early a stage, or too struggling a condition, the development of eggs in the *Psithyrus*, and, consequently, the murder of the *Bombus*, are delayed until the requisite degree of prosperity is reached.

The poor *Bombus* queen appears to have no chance of victory over the well-armed *Psithyrus*. I have taken a great many nests of *lapidarius* and *terrestris* in all stages, but have seen no evidence to show that the *Bombus* queen ever succeeds in killing the *Psithyrus*, or that she ever escapes being destroyed by the latter. Here, however, my observations are opposed to those of Hoffer, who found the *Psithyrus* queen and the *Bombus* queen living in the nest on good terms with one another, and both of them producing young males and queens. But Hoffer's observations were made chiefly on two species of *Psithyrus* which I have not been able to study, namely, *Ps. campestris*, which breeds in the nests of *B. agrorum* and *B. helferanus*, and *Ps. quadricolor*, which preys on *B. pratorum*. Evidently these species of *Bombus*, which, it may be noted, are milder tempered than *B. lapidarius* and *terrestris*, do not object, in Styria at least, to the *Psithyri* laying their eggs in their nests. Of course, the *Psithyri*, in these cases, do not rear so large a family.

I believe the *Psithyrus* queens do not kill one another, for I have never found a dead *Psithyrus* in a nest ruled by a *Psithyrus*. If several *Psithyri* find the same nest only one remains, although the others may make it their headquarters for a few days, as noticed in the above-mentioned nest of *B. lapidarius*.

As might be supposed, the *Psithyrus* is at first very prolific, but she ages and fails more quickly than a *Bombus* queen.

The eggs of *Ps. rupestris* are slightly longer, but much more slender, than those of *B. lapidarius*, being four times as long as they are thick : the eggs of *Ps. vestalis* are also slenderer than those of *B. terrestris*.

The development of *Psithyrus* through the larval and pupal stages is the same as that of *Bombus*, but the cocoons of all the species that I have observed (*rupestris, vestalis, distinctus,* and *barbutellus*) differ from those of their *Bombus* hosts in that they quickly lose their fresh lemon tint, and become of a dull ochreous colour, and later turn semi-transparent, crackling when they are dented—qualities that are possessed in only a very slight degree by the *Bombus* cocoons. The cocoons of *Ps. rupestris* form compact clusters like those of its host *B. lapidarius,* while those of *Ps. vestalis,* like those of *B. terrestris,* are only loosely attached to one another.

The *Psithyrus* kills the *Bombus* queen before she has laid the full number of worker eggs, consequently nests containing *Psithyri* are not very populous, the number of workers seldom exceeding eighty. In nests of *B. lapidarius* containing *Ps. rupestris* I have never known a *lapidarius* queen to be reared; similarly in nests of *B. terrestris* attacked by *Ps. vestalis* I have never seen a young *terrestris* queen.

In a *Psithyrus*-ridden nest a large number of the workers become fertile. In digging out a nest of *B. terrestris,* I can generally tell before reaching the nest that it has been victimised by *Ps. vestalis* if many workers containing eggs, known by their shining black abdomens with the yellow band on the second segment more or less obliterated, rush out of the hole to greet me. But it is a remarkable fact that as long as the *Psithyrus* queen reigns in the nest I have not known a *Bombus* male to be

produced, although in a nest containing *Ps. vestalis*
I have seen the *terrestris* workers lay their eggs.
In this same nest, however, I saw the *Psithyrus*
queen calmly eating the workers' eggs, and I think
that this is probably the way in which she always
disposes of them.

The *Psithyrus* queen pays close attention to her
new-laid eggs for several hours, giving the workers
no chance to molest them, but the workers soon get
reconciled to them, and henceforth they feed and
tend the *Psithyrus* brood with as much devotion as
if it were of their own species : indeed, they seem
sometimes to show a greater fondness for it. Hoffer
found that the larvæ of *Psithyrus campestris* died
in a case where the *Psithyrus* queen disappeared
while they were still very young, and concluded
that the *Psithyrus* contributes something essential
to the nourishment of her larvæ when they are
young. I have seen a queen of *Ps. distinctus*, a
species closely allied to *Ps. vestalis*, feeding her
young larvæ ; but some later larvæ of this queen
developed into fair-sized queens, although I removed
her when they were very small, namely, twenty-five
days before the queens emerged.

The hard, convex, narrow and naked underside
of the abdomen of the *Psithyrus* queen is unsuited
for incubating, and I have never seen her spreading
herself over the brood.

On July 14, 1911, I took a nest of *B. hortorum*
containing its parasite *Psithyrus barbutellus*. There
were 49 *hortorum* workers, many with shiny

abdomens, 16 young *barbutellus* queens, 2 small *barbutellus* males, and the body of the old *barbutellus* queen. The brood consisted of 38 *Psithyrus* cocoons containing pupæ, chiefly queens, and a cluster of 5 *hortorum* cocoons containing younger pupæ, all of which developed into males, besides a few larvæ and eggs. Probably therefore *Ps. barbutellus* is parasitic in the same deadly way as *Ps. rupestris* and *vestalis*.

There is a greater or less tendency in all the species of *Psithyrus* to resemble their particular hosts in the pattern and colour of their coat. It can hardly be doubted that this is of some advantage to the *Psithyrus*.

The origin of *Psithyrus*, more especially of its peculiar parasitical instincts, is an interesting question. If a specimen of *Psithyrus* be compared with a specimen of *Bombus* it is seen that the resemblance is not merely superficial but extends to nearly all the important details of structure, so that it is impossible to avoid the conclusion that *Psithyrus* has sprung from *Bombus*, and this at quite a recent period in the history of life. Moreover, the *Bombi*—and this is particularly interesting—show parasitical tendencies leading to the parasitism of *Psithyrus*. We have seen (pages 55-58) how the *Bombus* queens may enter the nests of their own species and kill one another, and how, in the case of the twin species, *B. terrestris* and *lucorum*, *terrestris* has extended this habit so as to prey on *lucorum*, killing the *lucorum* queen and getting the *lucorum* workers to rear her young in practically same manner as the *Psithyri* prey on the *Bombi*. It is a remarkable fact that the sting of the *terrestris* queen differs from that of the *lucorum* queen and approaches that of *Psithyrus* in being somewhat stouter and more curved, and having its thickened basal portion more parallel-sided when viewed sideways than in *lucorum*. There is, however, no evidence to show that any species of *Psithyrus* has sprung from the particular species of *Bombus* on which it preys, such resemblances as it may show to it in coat-colour, etc., being pretty clearly attributable to mimicry or exposure to the same conditions of life, and not to ancestry.

The males of the *Psithyri* are very fond of drowsing on flowers, especially on the heads of the knap-weed; they are even more fragrant than the males of the *Bombi*. In their play, which is probably amorous, they hover over meadows and grassy slopes. I know two exposed grassy inclines, both facing north-west and having no trees near, where, every August, the males of *Ps. rupestris* may be seen flying about, skimming the grass, sometimes in considerable numbers. I have seen the males of *Ps. barbutellus* hovering over the grass in the same way in July, but in the vicinity of trees.

I have dug up the hibernating queens of *Ps. rupestris, vestalis, barbutellus*, and *campestris* from banks facing north-west, occupying little cavities in the ground, about two inches below the surface, exactly like the queens of *B. lapidarius* and other *Bombi*.

IV

PARASITES AND ENEMIES OF THE HUMBLE-BEE

MANY kinds of small animals, chiefly insects, are to be found in humble-bees' nests. Some of these are chance visitors, with no particular business there, but others are dependent in some way upon the humble-bees, and several belonging to this class are very injurious to them, devouring the larvæ and pupæ.

Of the latter kind one of the most destructive is the caterpillar of the humble-bee wax-moth (*Aphomia sociella*). It feeds upon the brood of the humble-bee and probably everything else that can be eaten in the nest. The full-sized caterpillars are 1 in. to $1\frac{1}{4}$ in. long, pale olive-green above and yellow beneath, with the head and first segment orange-brown above. They weave a loose web about themselves into which they can retreat, safe from molestation by the humble-bees. They are very active and can run backwards as well as forwards. A hundred of these caterpillars—a nest infested with them seldom contains fewer, sometimes many more—

will completely destroy a large comb in a few days, riddling it with their silk-lined tunnels and reducing it to an impenetrable, sponge-like mass of web and debris.   Upon this lump the humble-bees, deprived of all their cells and brood, sit listlessly, unable to help themselves.   The caterpillars, when they are full fed, creep out of the nest in a body and spin their long tough cocoons, laying them side by side in a bunch.   The cocoons form a dirty white ball

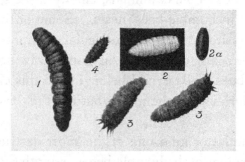

FIG. 15.—1, Caterpillar of *Aphomia sociella* ; 2, Larva of *Brachycoma devia* ; 2*a*, Puparium of ditto ; 3, 3, Larvæ of *Volucella bombylans* ; 4, Larva of *Fannia*.   All natural size.

which might easily be mistaken for a lump of rubbish. Here the caterpillars pass the winter, changing to pupæ in May.   The moths, which emerge in June, are dingy white, with brown markings and with the anterior wings tinged in front with green, which soon rubs off.   *B. derhamellus*, one of the surface-dwelling species that generally makes its nest earlier than the others, is very apt to fall a prey to wax-moth caterpillars.   I have also found them in the nests of *B. terrestris*, *B. hortorum*, and other under-ground dwellers, but I have never seen them in

nests of *B. lapidarius*. When the infested nest is
under the ground the full-fed caterpillars often spin
a web to the mouth of the tunnel and, climbing up
by this, spin their cocoons on the surface under a
stone or other protection.

Another insect that devours the brood is the
larva of *Brachycoma devia*, a two-winged fly. This
fly much resembles the common house-fly, but it
belongs to a different family, the *Tachininæ*, many
of the species of which, in their larval stage, are
parasitic in the bodies of various insects. The
larva of *Brachycoma* is of the shape usual in fly
maggots, namely, tapering to a point at the mouth
and truncated at the tail. It is white and trans-
lucent and bears no spines. Large specimens
attain a length of $\frac{5}{8}$ inch. Only about a dozen of
these maggots are usually found in a nest at a time,
but occasionally there are more, and I have seen the
brood in a nest of *B. pratorum* completely eaten up
by them. In the hot summer of 1911 the maggots
were seen in most of my outdoor nests. In a strong
nest of *terrestris* that I dug up, several of the queen
cocoons had a soft watery appearance and were
found when opened to contain the maggots ; there
were two, sometimes three, nearly full-sized maggots
in each cocoon with the shrivelled remains of the
Bombus larva, but there were no holes in the cocoons
so that the maggots must have allowed themselves
to be imprisoned with the *Bombus* larva when it was
spinning. I have seen the cocoons of *B. muscorum*
similarly occupied by the maggots, and find that it

is not always the larvæ in the largest cocoons that are victimised. The maggots usually creep into the nest material to pupate : the puparium is at first yellowish-brown, afterwards it becomes dark red. The perfect flies emerge in two or three weeks ; they may be easily obtained by placing the pupæ in a glass jar and covering it with muslin.

A very interesting dependant on the humble-bee is the large fly, *Volucella bombylans*. This handsome insect is about the size of a worker humble-bee and superficially resembles it closely, being of a stout build and covered with long hair which in some specimens is red at the tail and black over the rest of the body in imitation of *B. lapidarius* and *B. derhamellus*, and in others is white at the tail and marked with yellow on the thorax and base of the abdomen in imitation of the white-tailed, yellow-banded species, particularly *B. jonellus* and *B. hortorum*. Both forms occur plentifully in most districts and have, I believe, been bred from the same parent. But the mimicry of *Volucella* does not end here. It is very fond of sunning itself upon flowers frequented by humble-bees, such as the heads of the greater knap-weed (*Centaurea scabiosa*), and if a specimen be grasped in the fingers it makes a buzzing noise so exactly like a humble-bee that the fear of getting stung makes one feel inclined to drop it instantly. Moreover, these flies hover about over grassy banks, etc., in the same manner as a *Bombus*, searching for humble-bees' nests in which to lay their eggs, and at a distance I have often mistaken

one for a humble-bee. It is a singular fact that if
one kills a female *Volucella* she at once commences
to lay eggs, and so great is the vitality of the ovipositing apparatus that on one occasion I found
about a dozen eggs had been laid by a female in my
cyanide killing-bottle. No doubt if they are stung
to death by the humble-bees in their nests they are
always able to finish laying their eggs. The eggs
are large and hard, and when new laid, like those
of many insects, are thinly coated with a glutinous
substance, which quickly hardens and fastens them
to one another and to the object upon which they
are laid. The larvæ are of a dirty yellowish-white
colour and slightly flattened, and have a tough
wrinkled skin; they have six long spines arranged
in a semicircle under the anus, and also two rows of
rudimentary spines running the whole length of the
body on either side. They attain the large size of
$\frac{3}{4}$ in. to $\frac{7}{8}$ in. in length and $\frac{1}{4}$ in. in width. They
live in the debris under the comb and are said to be
scavengers; they certainly have done no harm to
the brood in my nests.

The larvæ of another kind of fly, *Fannia*, are
often found on the earth in the nest cavity. They
are considerably smaller than the *Volucella* larvæ;
their colour is dirty brown, and they have two
rows of spines running the whole length of the
body on either side and two more rows closer
together along the top; at the anal end the
spines are longer. These are also scavengers and
feed upon excrement. The perfect fly is dull

coloured and bears some slight resemblance to a house-fly.

Among the smaller and less important inhabitants of humble-bees' nests are beetles belonging to the genus *Antherophagus*. *A. nigricornis* is yellowish-brown, about $\frac{3}{16}$ in. long and $\frac{1}{16}$ in. wide.[1] Little flies with beautiful irridescent wings are sometimes to be seen running rapidly over the comb, the females with their bodies enormously distended : these are examples of *Phora vitripennis*. Possibly their larvæ consume the humble-bees' eggs.

The larva of the fly *Conops* lives inside the body of the larva and pupa of the humble-bee, the perfect fly emerging from the adult humble-bee sometimes after the latter has been killed and placed in a collection ; but I have not met with it in East Kent.

Neither have I found the rare ant-like fossor, *Mutilla europea*, in the nests, although it has been recorded from Hampshire in the nests of *B. agrorum* and several other species. The female of this remarkable parasite is about half-an-inch long, wingless and black, with the greater part of the thorax

[1] Other beetles that I have found in humble-bee's nests are *Antherophagus silaceus, Cryptophagus setulosus, Epuræa æstiva, Lathrimæum atrocephalum,* and *Choleva nigricans.* Probably the three latter were casual visitors.

A little *Braconid* (kind of ichneumon-fly), which Mr. Claude Morley has determined as being closely allied to (possibly identical with) *Histeromerus mystacinus*, was found in several nests containing *Antherophagus*, and is perhaps parasitic on it.

Mr. W. H. Tuck at Bury St. Edmunds found over fifty species of beetles and several Hemiptera, etc., in humble-bee nests, and also bred the following Diptera from larvæ found in the nests :—*Chrysotoxum festivum* (*B. lapidarius*), *Eristalis intricarius* (*B. agrorum*), and *Leucozonia lucorum* (*B. terrestris*). See Tuck's lists of species in the *Entomologist's Monthly Magazine*, July 1896, p. 153, and March 1897, p. 58.

red, and with three bands of golden hairs across the abdomen. The male is winged and has the abdominal bands silvery.

Almost every humble-bee's nest is more or less infested with mites, minute eight-legged animals related to the spiders and ticks. They are harmless to the humble-bees and feed on wax or on the food supplied to the humble-bee larvæ, for the young mites may sometimes be found swarming under their waxen coverings. When the young queens are developed these mites leave the comb and crawl on to their bodies, obtaining a lodgment among the hairs, and in this way they are carried into the new nests the following spring. Over a hundred of these mites may sometimes be found on a single queen ; they chiefly congregate on the back of the thorax and base of the abdomen.

Among the harmless denizens of the nest must be included a small lepidopterous caterpillar that much resembles the young larva of the wax-moth, but it is less active, spins no web, and does not attain a greater length than about a quarter-of-an inch.

A not uncommon parasite of the humble-bee is the thread-worm *Sphærularia bombi*. This worm has a remarkable life-history which is described as follows by Professor Sedgwick :—" There are small nematodes, the females of which alone are parasitic. These, after copulation in the free state with the small males, migrate into insects, and, under the favourable conditions of parasitism, not only increase enormously in size, but also undergo structural modifications favourable for the production of a large number of embryos. In *Sphærularia bombi*, the remarkable parasite of the humble-bee, the females, after copulating in the free state, migrate into the queen-bees that live through the winter. Here the gut degenerates and a kind of hernia of the body-wall, containing the generative organs, is

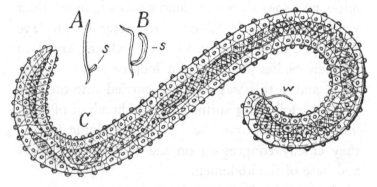

Fig. 16.—*Sphærularia bombi. A*, Female with partially protruded vagina, *s. B*, The same with still more developed uterine growth, *s. C*, Uterine out-growth fully formed containing ovary, oviduct, and uterus. *w*, The relatively minute body of the worm. All magnified about ten times. (After Sedgwick).

formed, while the body of the worm shrinks to a small appendage. The eggs develop in the body of the insect into larvæ which pass out of the body, become free and after some months become sexually mature." [1] The parasite lives in the abdomen and the uterine outgrowth, which may be found fully developed in the hibernating queens where it attains a length of about half-an-inch, looks at first sight as

[1] *A Student's Text-Book of Zoology*, by A. Sedgwick, F.R.S., vol. i. pages 282 and 283. See also R. Leuckart's *Neue Beitrage z. Kenntniss d. Nematoden*, Leipzig, 1887.

if it might be a portion of the intestine of the bee. I once found on a flower in the beginning of July an ailing old *terrestris* queen with her abdomen full of hair-like worms, probably *Sphærularia bombi.* The parasite is said to be particularly common in Epping Forest.

Badgers are said to be fond of scratching out and eating the nests. Moles and weasels also destroy them. But the greatest mammalian enemies of the humble-bees are shrews and field mice. These destroy the nests before any workers have emerged, devouring the brood, and they are the only verte-brates against which there is strong evidence of having destroyed any of the nests that I have kept under observation at Ripple. My experiences of their depredations, and also of the harm done by ants, will be given later (pages 116-119).

Not many animals prey on the adult humble-bee. It is well known that the red-backed shrike brings them to its larder, impaling them on thorns, but birds in general avoid them. Hoffer found swallows and domestic fowls catching and eating humble-bees ; but in East Kent I have never seen the former take them, and the latter, I notice, are afraid of them. I have seen places where the hibernating queens have been picked out of the ground, probably by birds.

Saunders says that dead humble-bees are often found in numbers in a mutilated state under lime trees, and explains that they have been caught, after they have filled themselves with honey and become drowsy, by the great tit and possibly other birds.

The bird after catching the bee picks a hole in its
abdomen, enjoys the honey it has eaten, and then
drops the quivering body, which falls to the ground.
He once had the opportunity of seeing the slaughter
going on, and was able to detect the great tit as the
murderer.[1]

Lastly, man himself must be reckoned among the
enemies both of the humble-bee and its nest, and the
humble-bees share the distrust of him that is universal among wild animals.   One is reminded of
Titania's injunction to the fairies in the *Midsummer
Night's Dream* :—

> The honey-bags steal from the humble-bees ;
> And for night tapers crop their waxen thighs,
> And light them at the fiery glow-worm's eyes ;

and of the request that Bottom, later on in the play,
makes to Cobweb to " kill me a red-hipped humble-
bee, and, good monsieur, bring me the honey-bag."
It may be remarked that it is not necessary to kill
a humble-bee to obtain its honey, for it can be made
to disgorge it by pressure on the abdomen from
behind.

[1] *Wild Bees, Wasps, and Ants,* by E. Saunders. (Routledge, 1907.)

V

FINDING AND TAKING NESTS

It was one of the pleasures of my boyhood, and it is no less enjoyable now, to find nests of humble-bees, especially of the rarer species, and transfer them to suitable places in the garden where their working can be watched. The operation of digging out the nest is full of excitement and surprises, and the humble-bees settle down quickly and happily in their new quarters, where, rid of parasites, they often succeed better than if they had been left alone.

A nest may be discovered accidentally, the bees having been noticed passing in and out; or we may specially go out to search for the nests, and this in itself is good sport. June and July are the best months in which to look for humble-bees' nests; but a few in an early stage may sometimes be found at the end of May, or we may take them in August after they have reached the height of their prosperity, although this is not so satisfactory.

It is important to choose a day on which there is little or no wind. The most promising places to search over are grassy banks and the borders of old woods, especially those containing decaying tree stumps. Other good places are the edges of meadows, paddocks, rickyards, rough waste ground, and ferneries consisting of grubbed-up tree-roots.

To discover the nests it is best to walk slowly along the foot of the bank or the outside of the wood, stopping at times, and all the while keeping one's eyes resting on a spot about twenty yards ahead on the bank or in the wood, ready to follow with the eye, if not on foot, every humble-bee that is seen or heard. Here the advantage of a calm day is realised, for the waving of the grass in the wind and the rustling of the foliage make it very difficult to see or hear any bee that is not very close, and impossible to keep it within sight longer than a moment, particularly as the wind may cause its flight to be somewhat erratic. If we are in East Kent, though other parts of the country can hardly differ much in this respect, we shall be rewarded within a few minutes by the sight of a humble-bee either leaving or entering its nest. The beginner, however, is very likely to be puzzled or deceived by the sight of a male of *B. pratorum* or *B. hortorum* lightly flying along the bank and pausing for a moment, without quite settling, at the foot of a certain tree, or in a shady recess under a particular shrub, followed sooner or later by another (see p. 13). But if we see a bee alight and not rise again we may be pretty sure we have found a nest, especially if its flight is heavy and deliberate. Care must be taken not to disarrange the grass

around the spot until we have ascertained the exact
place (which is likely to be more or less concealed)
where the bees go in and out, for the bees know
every stalk and blade about the entrance to their
nest, and if any of these are disturbed they will be
unable to show us the way in. We may now capture
a specimen to ascertain the species, and also find
out if the nest is underground or on the surface ; in
the latter case a gentle patting of the surrounding
herbage will elicit a muffled buzzing from it and so
reveal its position.   Should it not be desired to take
the nest at once, its position may be marked by
placing a stick in the ground close to it ; or, in the
case of a nest in an open bank, a stone or two from
the road or field may be placed opposite it.   Another
good way to remember the location of a nest is to
note two prominent objects, one much nearer than
the other that it brings into line, this imaginary line
being about at right angles to the bank or path
near which the nest is situated.

On calm days I have often discovered nests on
the roadside when out driving in my trap while it
is going at walking pace up or down hills.   An
excellent time to find the nests is in the evening
when the wind has died down, for the humble-
bees keep busy till just before sunset, and in warm
weather till dusk.

Many nests may often be discovered in newly-
mown hay-fields, where the bees, having been thrown
into confusion by the cutting of the hay, may be
seen hovering around the spots endeavouring to

find the way in. Unfortunately the hay-rake destroys many surface nests, consequently these should be taken directly the hay is cut.

As regards the taking of the nest, I have never been able to improve upon a method I devised when a boy.

The apparatus needed is quite simple—a strong trowel, two glass jars with narrow necks,—ordinary 1 lb. jam jars do very well,—and two squares of card large enough to cover the mouths of the jars; also a little box to hold the comb, though in my boyhood I used to carry it in my pocket-handkerchief.

The nest should be taken in the afternoon. Having found the hole in the ground that the bees pass through, I start digging it out with the trowel, taking care not to lose it. After going straight down for a few inches the hole generally takes an oblique, and finally an almost horizontal course. I dig a little deeper than the hole, thus keeping a cavity underneath it, into which all the loose earth falls, and out of which it can be easily shovelled. Sooner or later the hole may divide into two or more, and now it is a question which of these leads to the nest. One is not kept long in doubt. There is a rumbling, followed by an angry buzz, and out rushes a very fussy worker from one of the holes. She may tumble out into the cavity and there lie motionless on her back, ready to seize and sting anything that touches her, or she may attempt to seek safety in flight, or she may run back down the hole, growling as she goes. But before she has time to decide

which of these she will do I generally capture her
in one of the glass jars that I have in readiness,
clapping the cardboard over its mouth.   This jar I
stand in a convenient place on level ground about a
yard away and place a stone or lump of earth on
the cover to prevent the wind blowing it off.   Continuing to dig out the hole, it is not long before I
am greeted by another worker from the nest; she
is promptly captured in the other jar; this jar is
then placed, mouth downwards, on top of the first
jar, the two cards drawn out and the two jars given
a vigorous shake; this causes the bee in the upper
jar to drop into the lower jar containing her comrade ;
one of the cards is then quickly slipped over the
mouth of the lower jar and the stone replaced on it.
Thus all the bees that come out are caught, one by
one, in one of the jars, and collected in the other
jar.   The process of securing the bees in this way
is not so laborious as it may seem, and one soon
gets quite skilled and quick at it, so much so that it
becomes easy to catch two, sometimes three, bees at
a time.

Bees that lie on their backs will often grasp and
cling to a corner of the card if it be presented to
them.   The edge of the card may then be struck
sharply against the rim of the jar, with the result
that the bee falls into the jar.

The nest is generally a foot and a half to two
feet from the entrance and about fifteen inches deep,
but some are a yard or more from the entrance and
very deep.   To dig up deep nests a spade saves

time. To avoid waiting for bees to appear I find it
a good plan to blow down the hole; this, provided
the nest is not a great way off, or in a branch hole,
will bring up a bee, often several, immediately, or
at least it will cause buzzing, which assures one
that the right hole is being followed. For this
purpose a flexible pneumatic tube with a mouth-
piece would, I should think, be very convenient and
effective. As a boy I did not mind putting my head
down into the cavity to blow into the hole.

It is a good plan to stuff rags into any branch
holes one has temporarily abandoned, then one can
easily find them again and follow them up if the hole
that is being worked at proves abortive. In cases
where the hole is lost and choked the only thing to
do is to scrape out the cavity and leave the nest
until another day, when it will generally be found
that the bees have made a way for themselves in
and out again.

When workers begin rushing out in numbers in
quick succession it is a sign that we are not far from
the nest, especially if some of them are immature.
One must be careful not to plunge the trowel into
the comb, but generally one gets warning that the
nest is being approached—the loud deep buzz of
the agitated queen, often accompanied by the feeble
murmur of several recently emerged workers, which
are now likely to run out, and finally the appearance
of the nest material. In order to unearth the nest
without injuring it, it is advisable to undermine it.
While this work is proceeding the queen may rush

out, and one must take care to catch her without injuring her, and to put her with her children in the glass jar. She is generally easily secured, being often too heavy to fly, but should she take wing she will return sooner or later, though she will be rather wild and shy. If the colony is not in an advanced stage there will now be not a bee left in the nest, and one may boldly put in one's hand and lift it out. If, however, some males and queens have emerged, a few of these will be present, but they will be inoffensive, seeking only to hide themselves by creeping into the nest material, or to escape by flight. The comb should be carefully placed in the box with a little moss to keep it from rolling about, but without any of the nest material, which is likely to contain parasites or their eggs.

We may now turn our attention to the bees that were in the fields when we began operations, and which have been too shy to approach while we have been digging. If we step back a few paces they will gather round and, searching about, will soon find the nest material, impregnated with the odour of their nest. Alighting upon this, they may easily be caught.

Many people are deterred from taking a humble-bees' nest by the risk of stings; but really this is very small, and, if care is taken, may be ignored, except in the case of populous nests of *B. terrestris* and of *B. muscorum*, though with these only a little extra care and patience are needed. During the year 1911 I took nearly a hundred humble-bees' nests in all stages, including some strong ones of *B. terrestris*, and did not receive a single sting. The main points to be observed are to endeavour to catch every bee that comes out of the nest, and not to disturb the bees in the nest until nearly all have been captured. If a worker does escape now and then, it is not likely to attack you unless it is a *terrestris* or *muscorum*, and even these soon lose their aggressiveness unless several of their comrades have also been permitted to get away; the fugitives then rouse one another's anger, and if they pour out of the nest it is wise to beat an immediate retreat. The quickest way to take a populous *terrestris* nest is to stupefy the bees by stopping the hole with a rag on which have been placed a few drops of ether. In all cases the bees that return from the fields are perfectly harmless so long as they do not smell the nest, and if one keeps working at it they dare not approach near enough to do this. A queen humble-bee will never attack a human being, she only stings if she is held or crushed. When the workers are alarmed in their nest some of them crawl out and, turning over on their backs, lie motionless, perhaps for a minute or two, ready to seize and sting anything that touches them. The surface-dwelling species are specially given to doing this, often hiding themselves in the nest material, which is coloured like themselves, and one has to be careful not to put one's hand on them. The two surface-dwelling species, *B. sylvarum* and *B. helferanus*, are very calm when

their nest is opened; the majority of the workers
remain quietly on the comb, and the few that take
wing fly straight away.

The humble-bee does not lose its sting in the

FIG. 17.—Nest of *Bombus helferanus*, natural size.

act of stinging, like the honey-bee; it can therefore
use it more than once, but the poison is less virulent
than that of the honey-bee or the wasp.

And now, having all the bees safely collected in
one of the glass jars and the comb in the box, we

bring them home and prepare a domicile for them. The bees will be none the worse for being confined in the jar for one or two hours, provided they have been supplied with air and are not too numerous— fifty are enough for a 1 lb. jam jar of 2½ ins. diameter. The brood will not suffer from the temporary loss of bees if it is not exposed for long to hot sunshine or a cold wind.

Humble-bees cannot be kept in wooden hives like honey-bees, mainly for two reasons; one is that the jarring caused by the opening and shutting of the hive would throw them into excitement and confusion, and the other is that in order to keep the nest sweet and clean it is necessary for it to rest upon, or be surrounded by earth.

The simplest way to re-establish the colony is to place the comb, surrounded with a good supply of moss or curly, soft, dead blades of grass, upon the ground—any convenient part of the garden will do —putting a large box or flower-pot upside down over it to protect it from sunshine and rain. The bees will soon draw the material over the comb, making themselves very comfortable and warm underneath it. But if a view of the comb is desired it should be placed without any material under a Sladen cover (see page 109), or enclosed in a box scarcely larger than itself and having no top or bottom, the box being covered with a sheet of glass. This box may be placed on the ground inside an outer box, but for frequent or long-continued observations a much more convenient and satisfactory plan is to put it on a shelf in a wooden house specially fitted up to take humble-bees' nests, as explained in the next chapter.

Evening is the best time to put the bees into their new home. The comb having been placed where it is intended to remain, the bees, which have now become drowsy and listless, are shaken out of their jar on to it, the jar being sharply rapped so that they all fall out at once. When a flower-pot is used to cover the nest the jar may be simply placed upside down over the hole in the top of the pot. Finding themselves on the brood again, the bees revive and begin to clean and trim themselves, and in a wonderfully short time they recover completely, and run about over the comb carrying out their duties exactly as they did before the nest was taken. It is necessary at first to allow the bees no way of escape from the receptacle containing the comb, otherwise many of them would fly or crawl away and get lost ; but after an hour or two, when it is dark, and they have all settled down on the comb, a flight-hole may be made by raising the edge of the box or pot with a piece of wood or stone, or in any other convenient way, care being taken not to disturb the bees, and next morning they will be seen working busily, flying in and out in perfect contentment. If the old location of the nest is not far away some of the old bees will return to it, but not many, and their loss will soon be made good by the fresh ones emerging from the comb.

VI

A HUMBLE-BEE HOUSE

WHEN one is attending to and observing a number of colonies it is a great advantage to have them arranged close together, under cover, and at a convenient height from the ground. To meet this

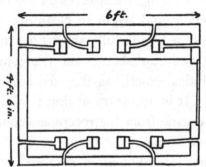

FIG. 18.—Plan of Sladen's Humble-bee House, showing position of holes for eight nests in the bottom of the house and of the nests on the shelves.

object I have devised a humble-bee house, which has enabled me to carry out my observations under the most favourable and pleasant conditions.

The house itself consists of an ordinary wooden hut, 6 ft. long, 4 ft. 6 in. wide, and 6 ft. high, covered with match boards. There is no floor. There is a window on one side, and this ought to be made to open. Inside the house are fitted two shelves, one on each side, extending from end to end of the house. Each shelf consists of a board 9 in. wide and 1 in. thick, and is at a height of $2\frac{1}{2}$ ft. from the ground. Each shelf will accommodate four colonies.

Figure 19 shows a vertical section through one of the domiciles. The comb is contained in wooden

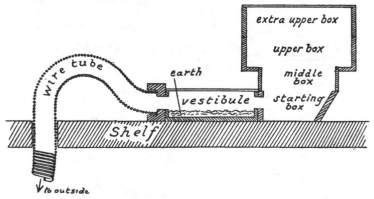

FIG. 19.—Vertical Section through a domicile in Sladen's Humble-bee House.

sections or storeys, which are multiplied as it grows. At the bottom is the starting section, measuring only $2\frac{1}{2}$ in. wide by $3\frac{1}{2}$ in. long on top and $2\frac{3}{4}$ in. long at the bottom, with an entrance hole at the end. This storey is large enough to hold the comb until the first batch of workers is succeeded by others. Then a second storey, measuring $3\frac{1}{2}$ in. by 4 in., is placed over the first, and this when filled with comb is capped by a third and still larger storey, measuring 4 in. by 5 in. Prosperous nests of *B. lapidarius* and *terrestris* usually require a fourth storey of the same size as the third. A sheet of glass is placed over the uppermost section, and through this the bees can be seen at work.

The domiciles are connected by flexible tubes made of spiral tinned-steel wire to holes bored through the bottom plates of the house. These holes are made as far from one another as possible, so that the workers of each nest do not mistake

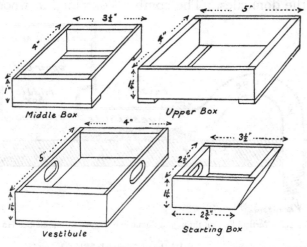

FIG. 20.—The Sections or Storeys shown separately.

their neighbours' entrances for their own—two holes being bored in each end and two in each side of the house. I found the workers were inclined to slip as they ascended the tubes, and in order to provide them with a good foothold, I intend next year to coat the tubes with varnish, and to pass sand through them while the varnish is wet.

The chief point requiring attention in an indoor nest of humble-bees is the keeping of it clean and sweet; but I find that this can be managed by having the nest sections no larger than the size of the comb, so that there are no large spaces in their corners where excrement can be deposited, and, at the same time, placing a vestibule, consisting of a box containing a little fine earth, between the nest and the entrance tube. The excrement is then ejected on to the earth, chiefly in the corners of the vestibule that are farthest from the nest; and at intervals the soiled earth may be removed and replaced by fresh earth. In this way a humble-bee's nest is not more difficult to keep clean than a bird-cage—in fact less so, for the earth need not be renewed oftener than once a week. The cleaning of the vestibule is only for one's own comfort, for the humble-bees do not suffer if it is left uncleaned. A sheet of glass covers the vestibule.

To prevent the humble-bees from lifting the sheets of glass covering the vestibule and nest, I place over each glass a piece of tile weighing about half-a-pound, with a square of felt under it. The tile and felt also help to conserve heat, and, in addition, keep out light, which the humble-bees dislike in their dwelling, although, like honey-bees, they soon become accustomed to it.

The best way to stock the house is to get the nests started out-of-doors under Sladen wooden covers (see p. 109), and then as soon as the first workers have emerged, to place them in the starting sections, feeding them for a few days until they can support themselves. Colonies of *lapidarius*, which is the species most likely to nest under the

H

wooden covers, do very well in the humble-bee
house, the nature of this species being often to
make its abode in this kind of situation. I have also
kept colonies of *terrestris*, *lucorum*, and *ruderatus*
in the humble-bee house, and they have done well.
Some of the colonies were dug up in the fields,
where they had a large comb and many workers. In
these cases the comb was packed into two 4-in. × 5-in.
sections, and no smaller sections were used. I see
no reason why colonies of the carder-bees should
not flourish equally well in a humble-bee house.

It is sometimes an advantage to be able to
separate the sections after they have been filled with
comb. This can be done by previously stretching
fine tinned wire three or four times across the
bottom of each section on fine nails driven into
the underside of the section. From one of my
lapidarius nests I removed the two bottom sections
filled with cocoons containing over four ounces of
thick honey without destroying the colony.

My humble-bee house was situated so that it was
shaded from sunshine during the greater part of
the day. This was rather important, for continuous
sunshine on the black felt roof would soon have
made the humble-bees uncomfortably hot.

The grass was allowed to grow high all round
the house. The workers found their way in and
out through it easily, and the variety of its growth
helped them to recognise the exact position of their
respective entrances.

When all the eight domiciles were occupied by

populous colonies, it was pretty to see the numerous workers busily flying in and out all round the house day after day, sometimes till late in the evening. It was interesting to watch them climbing the wire tubes, to follow them into their nests, and then to see them unburden themselves of their loads of pollen and nectar.

One great advantage in having so many nests under one's eye at the same time was that there was always something interesting going on in some of them. In one the queen might be giving the finishing touches to a cell in which she was about to lay a batch of eggs; in another the queen might be repulsing the attacks of the workers on her new-laid eggs; and in a third a worker might be seen injecting food into a wax-covered cluster of larvæ, or helping a baby brother or sister to creep out of its cocoon.

A very good time to watch the bees in their nests was at night, for while they were liable sometimes to be disconcerted by daylight, they completely ignored the light of a candle, provided they were not disturbed. At night, too, the full populations gave the colonies a very gay and animated appearance. It was a pleasant relaxation after the day's work to sit on a chair inside the house and study the various proceedings of the bees; and the stillness of night was conducive to close observation.

Except when I was making observations in the day-time I kept the window covered with a thick black blind. Without this the workers would have had a great deal of difficulty in learning how to find their way up and down the tubes.

A jar threw all the colonies into an uproar, the workers crowding together and tumbling over one another in their excitement; but they soon quieted down. Of course, as no bees could escape inside the house, there was no risk of getting stung. At first the least vibration caused a commotion; later on, however, the bees became accustomed to slight jars, and as a rule they took no notice of them, but occasionally, in some of the nests, the alarm would be raised by a worker or two, and it then quickly spread through the colony.

Frequent panics seemed to check the prosperity of the colonies: I therefore kept them as quiet as possible, and I made the door of the house to open noiselessly with a spring fastener so that my going in and out did not cause any disturbance.

It was necessary to remove the vestibule in order to clean it. To prevent the bees escaping when taking it away I slipped a strip of tin-plate over the hole in the bottom section. When I wished to confine the bees for a day or two, the tin-plate was slipped between the vestibule and the entrance tube. Sometimes I wanted to confine the queen and not the workers to the nest: this was done by substituting for the tin-plate a thin board containing a slot too small for the queen to escape, but large enough to permit the passage of the workers.

By taking care to avoid jarring and breathing upon the bees, I could always lift the glass off a *lapidarius* nest and inject honey into the cells, or withdraw honey from them, by means of a bulb-syringe, without fear of the workers becoming alarmed and flying out. I could also do this with weak colonies of *terrestris*, but populous colonies of *terrestris* threw themselves into a state of angry tumult as soon as I attempted to lift the glass.

Fig. 21.—Comb of *B. terrestris* from Sladen's Humble-bee House.

In populous nests, the bees fastened the sides of the glass to the top of the section with wax. If there was sufficient space they would also cover their comb with a dome of wax which, of course, hid the comb from view for the time being, but in *lapidarius* nests this wax was easily removed with a pair of forceps.

The bees generally built their honey-pots in the projecting wings of the sections. This is well shown in the accompanying photograph of a 4 in. × 5 in. section from a *terrestris* nest (Fig. 21).

VII

DOMESTICATION OF THE HUMBLE-BEE

IF one takes a colony with its brood and places it in a suitable receptacle in the garden or on a shelf of the humble-bee house, one sees how its existence is maintained and ends ; but one gets no information as to how the colony started and what took place during its early stages, a particularly interesting period. Nests in the stage before any workers have emerged may be discovered without much difficulty, and I find that it is possible to remove them successfully provided the operation is not carried out until after the first larvæ have spun their cocoons. For, while the queen has only eggs or young larvæ to care for, any alarm will cause her to desert the nest ; but as the brood grows older her attachment to it increases, and when it consists of pupæ that are not far from developing into workers she will not readily forsake it, any slight interference causing her to brood over it more devotedly than ever. At this period, if the queen is secured and placed in a box with the brood, she will sit on it until the workers emerge, provided of course that she is supplied with food. The queen's intelligence is seen at its best while she is thus caring for her brood, and her devotion to it, and her alertness on the slightest approach of danger, are most interesting to witness. She shows no desire to escape unless she is severely molested, and is quite content with her brood, anxiously incubating it day and night. It is important not to allow the queen her liberty, for if she is permitted to take wing she will return to the original location of her nest and, after vainly searching about for it there, she will fly away and be no more seen. As a precaution against loss, her wings may be clipped.

In May 1909 a *lucorum* queen was found to have occupied a mouse-nest under a large box that had been placed, bottom upwards, on the grass in my apiary the previous year. As the nest contained cocoons, I brought the queen with her brood indoors and fed her daily. She was devoted to her brood and never left it except to drink the honey I provided, or to perform other necessary duties, and she showed great agitation if, when returning to the brood, she did not find her way to it at once. When the workers emerged I clipped the queen's wings, and put the box containing the nest out of doors, under an inverted flower-pot, feeding the bees a little during the first few evenings. In a few days the little colony was able to support itself and it soon grew fairly prosperous. In due course it produced males and queens, but eventually wax-moth caterpillars devoured the brood.

ATTRACTING QUEENS TO OCCUPY ARTIFICIAL DOMICILES

Towards the end of April 1894 I made a ball of soft dead fibre from a tuft of pampas grass and placed it in some long grass. I left this ball undisturbed until June 4, when, on examining it, great was my pleasure to find it occupied by an *agrorum* queen. The material had been worked into a snug nest, inside which were five cocoons containing pupæ, one batch of very young larvæ and a honeypot full of honey.

The next spring I made about twenty nests from the soft dead blades of grass that are in some situations to be found in tufts under the new growth, this being the material of which the mice in the neighbourhood generally make their nests. In some cases I lined the nest with still softer material obtained by unravelling old rope and cutting it up into lengths of about half-an-inch. Although I placed these nests in the most likely places I could think of, such as on grassy banks and under ivy at the edge of a wood, not one of them was occupied by humble-bees.

No further attempts were made to lure queens with artificial nests until the year 1905, when it occurred to me to try to attract the underground-dwelling species by placing my nests under the ground.

Each domicile was prepared by cutting out a rectangular sod with a sharp spade; the floor of the excavation from which the sod was taken was levelled, and in the centre of it a small round cavity about 4-in. in diameter was made with a trowel. This cavity was to contain the nest. Next, a tunnel 1-in. in diameter and about 2 feet long, as a passage for the bees connecting the cavity with the surface of the ground, was made by driving through the ground a steel rod having

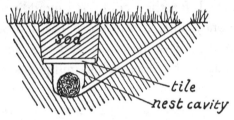

FIG. 22.—Section of Sladen's original device to attract underground-nesting humble-bees.

a thickened and pointed head. The mouth of the tunnel had the grass close around it plucked short so that it might be easily noticed by the searching queens. Most of the nests were made of dead grass like those made in 1895, but some consisted of soft moss torn to pieces and rolled into a ball, and a few were made of tow which, after it had been deodorised by exposure to the weather for a few days, was cut up into lengths of about half-an-inch. The nest having been deposited in the cavity, a tile was placed over it, the sod was put back in its place over the tile, and the edges of the sod were stamped down.

About twenty of these artificial underground domiciles were formed in May in and around my apiary, which is a favourite spot for humble-bees to nest in because the plot, almost an acre in size, is covered with grass that is never grazed, and is surrounded by young Austrian pines and other bushy trees, single specimens of which are scattered about.

An examination of these nests in the middle of June showed that several of them had been occupied by queens who had started breeding in them but had deserted, for they contained mouldy remains of pollen, honey-pots, and in some cases brood. The nests were more or less wet, and the nest cavities were occupied by centipedes, millipedes, slugs, and beetles. Some of the nests were saturated with water, and contained earth-worms and their casts. The cause of the queens deserting was apparently the damp and the vermin. In one of the nests, however, a poor *hortorum* queen was found dead with her feet entangled in tow, which, evidently, had not been cut short enough.

The following spring, in April, I put down forty more nests. The domiciles were made like those of the previous year, but I covered some of them with bee-hive roofs in the hope that these would keep them dry—an expectation that was not realised. The domiciles were numbered consecutively, the number being painted in black on a white stick, which was afterwards stuck in the sod covering each nest. I found these sticks very useful to mark the spots, which soon became overgrown with grass.

I also adopted the plan of examining all the domiciles every tenth day or so, replacing the nest material if wet, which was often the case, with dry material. The sods were lifted with an ordinary digging fork. When making these examinations I carried in my pocket a note-book, in which was recorded anything of interest that was noticed about the nests.

Nine out of the forty domiciles were tenanted by queens—three *lapidarius*, a *terrestris*, and a *hortorum* having been attracted in May, and two *lapidarius*, a *ruderatus*, and an unknown queen early in June ; but only three of the queens, namely, two *lapidarius* and the *ruderatus*, remained until their larvæ span their cocoons, and only one, a *lapidarius*, succeeded in hatching out her first workers and in establishing a colony. I formed the conclusion that the majority of the queens, after having occupied the nests and started to breed in them, found them so unsuited to their requirements that they abandoned them. The dampness of the nests or the vermin, or both, had driven them away.

During the next three years no further domiciles were prepared, but in 1909 two flourishing colonies with workers, one of *terrestris* and the other of *ruderatus* were found to be occupying two of the artificial domiciles made in 1906. Of course, by this time the ground had become perfectly hard, and the nests resembled natural nests. The nest material, though rotten and scanty, proved to be, on examination, not that which I had placed in the domiciles three years previously, but finer material brought there by mice, which had probably frequented the nests during the long interval, otherwise the tunnels could hardly have remained open so long.

The following winter, my interest in artificial domiciles having revived, I determined to give them a much more extensive and thorough trial than previously.

One of the chief difficulties in preparing a large number of domiciles is to get a sufficient supply of suitable material for the nests. My experience seemed to show that moss is not so suitable as fine, half-rotted grass, so I had some of the grass in the apiary mown in January, when much of it was dead, and in March when it was dry I stored it in empty hives, carefully picking out the stalks and heads, and cutting it with a pair of large scissors into lengths of about two inches. In this way I got an abundant supply of material, but it was far from being as fine as that of a mouse's nest, and I feel sure the results would have been better had the material been more suitable.

I placed about a dozen nests on the surface of the ground under bee-hive roofs, inverted boxes, and even tins. These nests kept fairly dry, but, partly I think because it is not the regular habit of any species of humble-bee to occupy nests in such positions, and partly because the material was so coarse, only one was occupied. This nest had been remade by a mouse, and therefore consisted of the fine material I had failed to supply. It was under a hive roof, and was found to be occupied by a *hortorum* queen on June 2. The queen was occasionally seen passing in and out up till June 9. The next day, the queen not having been observed, I lifted the roof and was sorry to find the nest disarranged, with no trace of comb. A weasel had

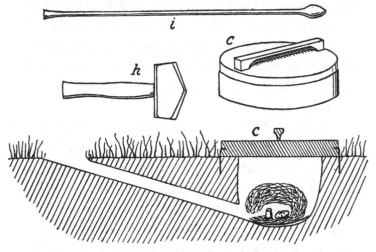

FIG. 23.—Sladen's Domicile for Humble-bees.
c, Sladen's wooden cover ; *i*, implement for making the tunnel ; *h*, mason's hammer for driving implement through the ground.

been seen five yards from the spot the day before ; in all probability the brood had been eaten either by this animal or by a shrew.

But my expectations for 1910 were centred in a new form of cover I had designed for underground domiciles, consisting of a thick wooden disc having around its edge a band of sheet metal projecting downwards. The metal band had its upper edge turned inwards at a tin-plate factory, and this

edge was inserted in a groove made in the edge of
the wooden disc, as shown in Fig. 23. This made
a splendid joint, remaining practically waterproof
and vermin-proof in all weathers. To prepare the
domicile, it was only necessary to place the cover
over a cavity made in the ground for the nest, and
to tread or stamp on it, so that the metal rim entered
the ground. Thus no crevice was left for vermin
or rain to get in. The cover was also easily lifted
and replaced. It seemed a great improvement on
the old method of digging out a sod and placing
a tile under it, and my hopes for it, as will be seen,
were justified.[1]

Between April 1 and the middle of May, seventy
nests were laid down under these covers in and
around the apiary, in my garden, and around the
edges of the surrounding plantations of trees. They
were placed in grassy ground where the turf was
firm and fine, the metal rim cutting into the turf and
making for itself a bed, out of which it could be
lifted without raising the surrounding surface when
I wished to examine the nests. The nests were
made of the prepared grass already described. The
tunnels connecting the nests to the surface varied in
length from about 15 in. to 18 in., and were at
first made about $\frac{3}{4}$ in. wide, but, as many got choked,
they were enlarged about the middle of May to
$1\frac{1}{8}$ inch.

The nests under these close-fitting covers were

[1] This cover has been registered, and arrangements have been made for
dealers in bee-appliances and entomological apparatus to supply it.

not altogether free from slugs, centipedes, etc., nor
did they keep perfectly dry, so I still found it
necessary to replace them with dry ones, and to
clear the tunnels, every tenth day or so; but this
was because the weather was showery, for in July,
when the ground got dry and hard, the nests
remained dry and the vermin left them.

I found that the chief cause of the nests getting
wet was contact with the damp earth. By taking
care that they did not touch the sides of the cavity,
and by placing discs of tin-plate under them, I was
able to keep them much drier.

As slugs and centipedes were often found and
killed in the tunnel when I cleared it, it was evident
that they got into the nest-cavity by passing through
the tunnel; I think, however, that some of them
travelled through the soil near the surface. The
slugs wetted the nests by crawling over them,
and so helped to draw moisture into them.

Although queens were less plentiful than usual,
owing to the previous unfavourable season, and
although fourteen out of the seventy nests laid down
were rendered useless because they were occupied
by ants or washed out by floods caused by thunder-
storms, twenty-one[1] of them were occupied by queens,
and in nine[2] the larvæ span their cocoons. These
results showed that the queens much preferred my
new wood-covered domiciles to the sod-and-tile
covered kind.

[1] 13 *lapidarius*, 3 *latreillellus*, 1 *terrestris*, 1 *ruderatus*, 1 *hortorum*, 1
sylvarum, and 1 species unknown.
[2] 5 *lapidarius*, 2 *latreillellus*, 1 *hortorum*, and 1 *sylvarum*.

It seemed that the only way to further escape from moisture and vermin was to go back to the plan of placing the nests deep in the ground and providing them with a long tunnel, as in the 1906 experiments, but to line the sides and top of the nest-cavity with some kind of material impervious to moisture.

This end I endeavoured to accomplish with a device that I shall call a "tin domicile," because its essential part was a cylinder of tin-plate, open at both ends, about 5 in. in diameter, and 5 in. or 6 in. high. My first experiments were made with ordinary tins, which, though not very suitable, were good enough to test the principle. I first removed the bottoms of the tins with a pair of strong scissors. To make the domicile, a cavity about 15 in. deep and a little wider than the size of the tin was dug in the selected spot. While I was digging this cavity with a trowel, my assistant made the tunnel or passage for the bees, which was about 2 ft. 6 in. long and $1\frac{1}{8}$ in. wide, by driving a long steel implement through the ground with a mason's hammer, directing it so as to emerge at the bottom of the cavity. If the head passed below the cavity the cavity was deepened to reach it. While the head of the implement was still in the cavity the tin, having its lid on, was placed in the cavity over it. The space outside the tin was then filled up with earth to a height of about 2 in. below the top of the tin, the earth being rammed down tight. After this the implement was withdrawn, the lid of the tin

having been previously lifted off, so that any loose
earth that might drop into the tunnel could be seen
and removed. Finally the nest was placed inside
the tin, the tin closed, and a board placed on the
surface of the ground over the spot.

Rust soon caused the lids to stick to the cylinders,
so I had to substitute lids of a larger size. To prevent ants and other vermin from getting into the
nest-cavity through the joint round the lid, I placed
a sheet of felt between the lid and the cylinder.

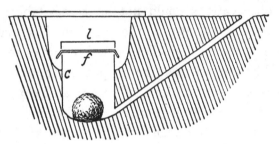

FIG. 24.—Vertical Section of Sladen's Tin Domicile.
c, Tin cylinder ; *l*, tin lid ; *f*, felt.

The first eight tin domiciles were put down on
May 23. Unfortunately these were provided with
small flat holes, $\frac{3}{8}$ in. by $1\frac{1}{2}$ in., some of which must
have got choked very soon, probably by worms
travelling across them ; nevertheless, two of them
were occupied by queens, both being *lapidarius*.

Seven more tin domiciles were put down on
May 27 with round holes $1\frac{1}{8}$ in. wide. Every one
of these became tenanted, six by *lapidarius* queens
and one by a *latreillellus* queen.

Finally, four more were put down on May 30
with $1\frac{1}{8}$ in. holes. Two of these became wet after

I

a thunderstorm, because the mouths of the holes were in a concavity; but of the two remaining, one was occupied by a *latreillellus* queen.

Of these ten queens, all except one succeeded in rearing her brood as far as the cocoon stage. But from this stage onwards the colonies did not thrive nearly so well in the tin domiciles as in the shallow, earth-walled domiciles under the wooden covers. The tin domiciles were not so sanitary, and another drawback was that they were so deep in the ground that observations could not be comfortably made. The greater success of the tin domiciles than the wooden-covered domiciles in attracting queens was probably due to their being put down at a more favourable time and in a more favourable place, for in the following year (*vide infra*) they did not prove so attractive to queens as the latter kind, which would have been more successful in 1910 if the season had not been abnormally damp.

Unfortunately many of my small tenants came to grief; but, thanks to my being able frequently to examine the nests, the cause of failure was ascertained in almost every case, and light was shed on how humble-bees fall a prey to enemies and adverse conditions. Moreover, my experience in attending to the nests enabled me to watch in detail, step by step, exactly how the queens proceed in nature to establish their homes, and thus most of the information about this given in the Life-History was obtained. I paid as much attention to my protégées as time permitted, and many of them would have been saved had I understood better how to protect them; but the dangers to which they were exposed were often not realised, and the way to combat them not discovered and put into practice until too late.

As the queen humble-bee works for nearly a month in all weathers, flying at all times of the day from early morning until dusk, I quite expected a considerable proportion of my queens to get lost, but I was pleased to find my fears almost groundless, for in every case in which a queen possessing brood disappeared, except one, it was found that she had deserted her nest because of some accident to the brood. Indeed, humble-bee queens seem to bear a charmed life. Undoubtedly they are avoided by the majority of insect-feeding birds and animals, which must recognise them by their massive appearance, their striking colours, and, when flying, by their loud peculiar hum. The scent of the humble-bee is disliked by many animals, their instinct leading them to flee from it; as may be seen, for instance, if one presents a humble-bee to a dog, who will turn away in disgust, although he will readily snap at a fly.

But there is no charm about the life of the humble-bee's brood, the eggs being very much relished by ants, and the larvæ and pupæ by mice and shrews. None of these enemies, so far as I have been able to observe, dares to attack the brood when the queen is in the nest. Here, if the queen is disturbed in the least, she warns the foe with loud angry buzzes, which are kept up for some time, and if further alarmed she will rush about and make a great fuss. In this connection it is interesting to note that the nature of the queen is always to remain at home, except when she is actually gathering food. Even if, through stress of weather, she is starving, she remains sitting faithfully on her brood, cold and lethargic, but ready to wake up and buzz should any intruder approach.

In three cases it was ascertained that the queen deserted her nest because her first eggs were devoured by ants, the offenders being in two instances the common black garden ant, *Lasius niger*, and in one instance the large brown ant, *Myrmica rubra*. These two species of ants were very common, as was also the yellow garden ant, *Lasius flavus*, but this was not observed to do any mischief. In all three cases the eggs were eaten, and all hope of saving the nest gone, before I noticed anything wrong; but in another nest under a wooden cover I had the good fortune to discover the ants before they had begun to attack the brood, and succeeded in protecting it from them, a good colony resulting. The method I employed was first to destroy every ant I could find in the nest and tunnel, and then to make a shallow trench around the wooden cover and entrance of the hole, pouring into the trench a mixture of turpentine and paraffin oil, the strong smell of which acted as a deterrent to the ants. Of course the smell gradually grew weaker, and I renewed it five days later by pouring some more paraffin oil, which evaporates less rapidly than turpentine, into the trench. Further particulars of this case are given on p. 229. For readers who may employ this remedy, I may remark that it is advisable to apply the oil at night in order to give it time to soak away before the queen flies again, because contact with the liquid oil would kill her.

This operation requires too much attention to be employed merely as a precaution against ants, and it seems that the only way to avoid the possibility of destruction by ants is to place the nests at a safe distance from the ants' nests. Unfortunately *Lasius niger* is abundant almost everywhere, living in the ground in small colonies, of which there is no indication on the surface. These colonies mostly dwell near the surface, but I have found them at depths extending to eight inches. One of my nests destroyed by *Lasius niger* was in a tin domicile, which, being at a greater depth, and having a longer tunnel, ought, one would think, to have afforded greater protection from ants than the wooden-covered nests situated only just beneath the surface, and provided with only a short hole. Ants are more active in damp seasons than in dry ones, and in the dry season of 1911 none of my nests were destroyed by them. Fortunately ants can do no harm to brood sufficiently protected by bees, as is shown by the fact that, later in the season, a colony of *Lasius niger* took up its abode under the wooden cover of one of my strong colonies of *lapidarius*, and the two colonies lived for some time together, the ants eventually disappearing without having harmed the bees.

Several more of my nests came to grief through a sudden and mysterious destruction of the brood. In three or four instances the brood was in an advanced stage, within a few days of the emergence of the workers. The cocoons were torn open, and the fragments were scattered about the nest, which was in a state of great disorder. I never succeeded in catching the assailants at work, but I set traps in two of the nests the day after the cocoons had been rifled, baiting them with the empty cocoons, and caught shrews which I have no doubt were the culprits, as these animals are insectivorous, and the brood of a humble-bee would make a dainty meal for them. In another case I found a field-mouse (*Mus sylvaticus*) occupying a nest from which the cocoons had disappeared.[1] Probably the common house-mouse also eats the brood when it can get a chance. Field voles (*Arvicola agrestis*), distinguishable from the true mice by their shorter tails, stouter bodies, and smaller eyes and ears, were common in the neighbourhood of my nests, and were sometimes found occupying the empty ones, but there was no evidence that they attacked the brood. Some of the nests had their material added to, others were largely reconstructed, and I think this must have been chiefly the work of voles.

[1] I cannot corroborate Col. Newman's statement, quoted by Darwin, that humble-bees' nests are more numerous near small towns than in the open country, believed to be because the cats in towns keep down the field mice, and think that this must be the case only in particular localities.

The nests destroyed by the shrews and mice were promising ones, and I quickly set to work to devise means to exclude these small mammals from the nests that they had not yet attacked. A shrew is really a very small creature, and, although it looks much larger than a humble-bee, its skeleton being internal, it can squeeze through almost as small an aperture as a queen humble-bee, whose skeleton is external and unyielding. I found by experiment that a *lapidarius* queen could comfortably pass through a square wooden hole measuring $\frac{9}{32}$ in. × $\frac{7}{16}$ in., and about $\frac{3}{8}$ in. long, but I had so much difficulty in pulling a small shrew through it that I do not think it could possibly have managed to creep through by itself. Covers containing holes of these dimensions were therefore placed over all my *lapidarius* nests. The cover consisted of a small tin lid with a large hole punched in the middle. Over the hole were fixed two strips of wood, one on either side of the hole, one of the strips being fixed to the tin, and the other strip being connected to the first strip by a pair of screws, which made it possible to vary to a nicety the size of the aperture between them, small blocks of wood of the width of the aperture required being placed between the strips, inside the screws. These covers were pressed into the ground over the mouths of the tunnels leading to the nests. The queens showed great intelligence in quickly recognising the holes in the covers, and seemed to enjoy passing through the narrow apertures. Unfortunately the queens at first had a difficulty in obtaining access to the holes from the inside, and to this defect, which was afterwards put right by filling the inside of the lid with clay, I owed the loss of three mothers of families.   No brood was injured or destroyed in the nests protected by these mouse excluders, but by the time they were placed in position most of the nests had almost passed through the period of danger, for I have never seen a nest containing workers injured by a mouse.

No doubt there are other animals that may destroy the brood of the humble-bee. Slugs and millipedes are vegetable feeders, and I have no evidence that they did any direct harm to the brood in my nests ; indeed, I think they never entered the nests as long as the latter remained fairly dry. But the common centipede (*Lithobius*) I regarded with considerable suspicion, for it is carnivorous, and also very active. Several times I found these centipedes in the cavities containing nests occupied by my queens, and once I saw one biting some cocoons I had lifted out of a deserted nest and had left lying on the ground for half an hour.

To the above causes of loss I must regretfully add another—my own blundering.

It has been mentioned that the queens are liable to forsake their nests if they are disturbed in them, and that very little interference causes a queen to desert her brood when it is young, but that after the cocoons are spun she becomes more attached to it, and when the workers are about to emerge only a severe fright will cause her to abandon it. A fright, however, will not lead to desertion if the queen can be kept from flying, and can be induced to settle down again to cherish her brood—not an easy matter when the brood is very young.

I had learnt these principles from the experience of former years, and they were carefully observed in my manipulations. I found the best time to open a domicile, should the queen be at home, was at dusk, for then she was disinclined to fly, and though she might be much excited she would soon settle down, and by morning, so far as one could tell, she had completely forgotten the incident. But I found it inadvisable and almost always unnecessary to open the domicile when the queen was at home. Such operations as destroying vermin, renewing nest material, and examining the brood, were carried out in full daylight when the queen was out. If a queen approached while I was doing something to her nest, I generally had time to put the nest back and close down the cover before she alighted, but rather than let her enter the cavity while it was still open and so discover, or even suspect, that her nest was being tampered with, I did not hesitate to frighten her off, although I often had considerable difficulty in doing so if the young were about to emerge. I often had to wait some time before seeing the queen depart from her nest, and, being busy, I used to ascertain if a queen was at home or not by rapping on the wooden cover, or, in the case of tin domiciles, on the tin lid. If she was there she would buzz. The rapping did not in any case cause desertion ;

on the contrary, some of my queens became so accustomed to frequent rapping that they failed to answer unless I knocked very hard. Rapping at night, when the queen ought to be at home, was a ready means of ascertaining if all was well.

Considering the frequency of my examinations and manipulations, and the various delicate operations carried out for the first time, freedom from accident could not have been expected. That the mishaps were so few shows how well suited humble-bees are to treatment, especially when it is remembered that with manipulations on honey-bees smoke is always used as a quieter. Smoke has no effect whatever upon humble-bees.

Accidents sometimes resulted from the queen staying from home for a day or two. Imagining she had got lost, I removed the brood to save it, and when the queen returned later, finding her brood gone, she deserted.

Two losses in tin domiciles were due to another kind of blunder. Lifting out the nests and examining them caused them to expand so much that when they were put back they practically filled the tins. The poor queens on returning home were unable to find their brood in the disarranged mass of material and, losing heart, deserted. The remedy, of course, was to use less nest material or a larger tin.

Several incidents showed how careful one should be, after making an examination, to leave the brood in a position where it will be readily found by the queen on her return to the nest. In the course of a week or two the nest subsides and adheres more or less to the ground : when it is put back carelessly into the cavity there may be a space left under it, into which the queen on her return is very likely to run. Here she may remain searching for her brood and fail to find it, though it be only just above her. Inability to find her brood soon after she enters the nest fills a queen with excitement and fear, and leads to desertion.

However much I might alter a nest or interfere with the brood in the queen's absence, she seemed never to notice it on her return nor to recognise the smell of human hands. In two nests of *lapidarius*, I opened the first cell of eggs to see how it was constructed, but the queens on returning home repaired the damage, and both nests developed into colonies.

The weather was favourable almost every day up to June 24, but from that date until the end of July it was exceptionally cold and cloudy for the time of year, favourable days being scarce and many days so stormy that very little food could be gathered. At the time the bad weather commenced most of the queens were getting past outdoor work, and the labour of providing food devolved upon the workers, which in nearly all the nests proved unable to supply the daily wants, so that I had to feed the little colonies regularly to save them from perishing.

Occasional periods of semi-starvation, lasting for a day or two, do no harm to a colony of humble-bees ; the bees simply become drowsy, remaining in

a state of suspended animation, until there comes a
fine morning with a rising temperature, when they
wake up and resume work with undiminished energy.
But if starvation is long continued the comb grows
mouldy and bees and brood die.

Experiments with humble-bees in previous years
had elicited the interesting fact that feeding them
tends to make them lazy, although feeding honey-
bees has no such effect. The reason may perhaps
be that the humble-bee worker, being less differenti-
ated from the queen than the honey-bee worker,
follows to some extent her mother's instinct of
ceasing to work as soon as she finds food provided.
To prevent my humble-bees from becoming lazy I
never commenced to feed a colony until it became
quite necessary, and then I fed it only in the evening
when the day's work was done, except in bad cases,
which were fed also in the morning. In continuously unfavourable weather I found that the colonies
that were well and regularly fed always did better
than those that were fed only occasionally.

I find two parts of honey mixed with one of
water makes a suitable food—or rather drink—for
humble-bees. But if this mixture is left unconsumed
from day to day it soon sours, and I usually add to
it some of the syrup I make for my honey-bees,
because this contains a little naphthol-beta, a germi-
cidal drug which prevents fermentation.

The food is injected into the cells by means of a
bulb syringe such as is used for filling fountain pens,
procurable from any stationer for 2d. For occasional

feeding, the honey-pot or some of the empty cocoons are filled ; but if a regular supply of food is likely to be needed I find it much better to provide one or two large artificial honey-pots, made of bees-wax by dipping several times into molten bees-wax the rounded end of a wooden stick previously moistened with water. The artificial pots need to have their bases firmly fixed to a piece of sacking, or to a mat of nest material by means of melted bees-wax, otherwise the bees will detach and capsize them.

In endeavouring to help on a struggling colony I often gave it a few just-emerged workers from a more prosperous one that could spare them.

The weather growing extremely bad, I found that many of my colonies were losing workers faster than they emerged from their cocoons. Despite all the feeding and attention I could give them they failed to make progress. Some of the colonies that had been fed irregularly were reduced to three or four small lazy workers sitting with the queen on a lump of mouldy, undersized cocoons containing diminutive pupæ whose development was being checked and delayed by repeated chills. It was plain that such colonies were on the verge of perishing, and that, even had the weather improved at once, their survival would have been doubtful and their progress at best very slow, for the queen was too worn to work much, and during the next three weeks only a few puny bees could have emerged. For a colony to prosper there must be a good work-ing force of strong young bees, constantly reinforced by fresh ones emerging : this can only happen if there is plenty of brood in all stages.

I found it possible to save these perishing colonies by bringing them indoors, where I confined each colony in a box and fed it liberally with honey and also with pollen that I obtained from a honey-bees' comb and placed in a special cell. This treatment, continued for about a fortnight in the case of the most impoverished colonies and for about a week in those that were stronger, produced in every case a wonderful improvement. The bees appeared to take fresh heart, tending the brood with increased zeal, so that the pupæ developed into bees without any further delay ; the half-starved larvæ rapidly grew to full size, span their cocoons, and later on developed into strong and vigorous workers ; the queen also recommenced laying eggs. In short, prosperity returned, and the colonies when put back into their domiciles were generally able, after a few evenings' feeding, to support themselves. Moreover, it was much easier for me to look after the colonies indoors than in their respective domiciles. One poor *latreillellus* queen and her young were nursed back to life in this way from the last stage of exhaustion. I unexpectedly found her in one of my nests when lifting the unoccupied ones as late as July 27. She was in a drowsy state and had a little lump of brood, most of which proved to be dead, and one drowsy worker. I gave them a worker from another nest, and after about two weeks' treatment they recovered.

In some nests I tried a longer course of treatment, but found that after about three weeks' confinement, sometimes less, the colony ceased to thrive, and signs of degeneration began to appear, the larvæ failing to grow to full size and delaying to spin their cocoons. The particular cause of this trouble was not ascertained, but it may be that the larval food contains something that is gathered in the fields besides honey and pollen, or that the workers are unable to prepare this food properly without exercise. It was noticed that a small colony throve better and longer in confinement than a populous one, and that the pollen-storing species *lapidarius* and *terrestris* were less quickly affected than the pocket-maker, *latreillellus*. A weak colony of *sylvarum* was confined twice with good results.

In the domiciles occupied by *lapidarius* colonies, as soon as the workers became busy and the queens ceased to fly I reduced the size of the hole in the mouse-excluder so that it would also exclude the parasite *Psithyrus rupestris*. To deter wax-moth and *Brachycoma* I relied upon balls of naphthaline, placing two or three of these in the grass close to the mouth of the hole and eight or ten around the wooden cover, but I found that these did not keep out the little *Phora vitripennis*.

I was pleased to find that the loss of nests in my artificial domiciles, though great, was, so far as I could ascertain, much less than that which occurred in nature. In the proximity of my domiciles I had discovered five nests commencing in natural holes, having in each case seen the queen flying in and out on several different occasions. Each was of a different species—*lapidarius, terrestris, ruderatus, hortorum,* and *latreillellus*—and I was looking forward with pleasure to digging up the nests and transferring them to observation domiciles as soon as the first workers appeared. But no workers were seen, except in the case of the *latreillellus,* and here when I came to take the nest the queen was not to be found.

The weather improved towards the end of July, and in August nearly all my nests under wooden covers developed into populous colonies.

The experiments were continued on a small scale in 1911. This season was in many ways the antithesis of the two preceding ones, the weather being continuously dry and warm, with the exception of a short period of showers and coolness towards the end of June, which, however, did not cause the humble-bees any inconvenience. Unfortunately the previous bad season had rendered queens less plentiful than usual, and *lapidarius* queens in particular were scarcer than I had ever before remembered them to be. Nine domiciles were made with Sladen wooden covers : three of these were occupied, two by *lapidarius* and one by a *latreillellus.* Twenty-two tin domiciles were made, but only three were occupied, two by *lapidarius* and one by *sylvarum.* Twelve domiciles of terra-cotta, constructed on the principle of the tin domicile, but with the diameter at the bottom much larger than at the top, were also

laid down : three of these were occupied by *lapidarius*
and one by *terrestris*. All these proved successful,
except one in a terra-cotta domicile, and, without
any aid from me, developed into colonies, although
the colonies in the tin and terra-cotta domiciles did
not flourish so well as those under the wooden covers.
As already stated, no damage was done by ants.
Neither did mice or shrews destroy any nests. To
keep these out, as soon as I saw a queen going in
and out of a nest, I pressed into the ground around
the mouth of the tunnel a tin cylinder about 5 inches
high and 5 inches in diameter, and open at both ends.
These cylinders, I believe, afford greater protection,
with less risk, than the mouse-excluders used in
1910; also, they probably give a certain amount of
protection against ants. The queens soon learnt to
fly in and out of them.

The fine dead grass used as nest material in 1911
was of two kinds, both of them superior to and more
easily obtained than that of 1910. The better kind
was scratched up by fowls which were confined to
a run which was moved once every day about a
pasture of fine grass in February, before the new
season's growth had begun : this material had the
great advantage of being curly, and it required
very little picking over and snipping up to prepare
it for use. The other kind was raked out of tufts
of fine grass in March and April.

Reviewing all my attempts at making domiciles
for queens to occupy, I consider that much the most
promising were those under the wooden covers, and

K

to readers who may like to repeat my experiments and watch humble-bees starting their nests in their gardens, I would recommend the construction in April and May of domiciles with these covers, as shown in Fig. 23. As long as the weather remains fairly favourable the starting colonies will need no attention, except perhaps the placing of cylinders of tin over the mouths of the holes to protect them from mice and shrews; but if they are to be tided over a long-continued spell of unfavourable weather in June or the early part of July, it will be necessary to bring them indoors and feed them.

Of all the queens that started nesting in my artificial underground domiciles during the three years 1906, 1910, and 1911, sixty-six per cent were *lapidarius*, twelve per cent were *latreillellus*, six per cent *terrestris*, and four per cent of each of the following species—*ruderatus*, *hortorum*, and *sylvarum*. The great preponderance of *lapidarius* over the other species was remarkable, the number of *lapidarius* being greater than all the rest put together; and though, no doubt, it was partly due to the fact that *lapidarius* appears late, when natural holes are overgrown, and therefore difficult to find, it is probable that an artificial domicile is more to the liking of *lapidarius* than of the other species.

GETTING QUEENS TO BREED IN CONFINEMENT

My first attempt at getting queens to start breeding in confinement was made when I was a boy. I caught a *terrestris* queen that was searching for a nest, and confined her in a box in which I had placed an artificial nest, keeping her supplied with food. She showed several signs of preparing for a family ; for instance, she would sit in the nest a good deal, and was careful not to soil it, and also she made a great fuss when disturbed. But later these signs disappeared, and she became restless.

I repeated the experiment with a good many *terrestris* queens : some gave the same result, others took no interest whatever in the nest.

I next tried placing two *terrestris* queens instead of one in each box. With these I got much better results, and many of them went so far as to lay eggs. Although the queens were often afraid of one another when first put together, they soon became friendly, provided they were allowed to have sufficient room, and the eggs were usually laid about a week after confinement had commenced. But unfortunately, about the time that the eggs were laid, sometimes a day or two before, sometimes a day or two after, one of the queens always killed her companion. The surviving queen at first paid every attention to her brood, but within a week she always deserted it, although I took the greatest care not to disturb her, and supplied her with plenty of food, both honey and pollen. In several cases, however, it was ascertained not only that the larvæ had hatched, but that they had been fed by the queen, and had begun to grow.

Hoping that the queen's affection for her brood

might have been maintained if I allowed her to fly, I made a hole in one of my boxes and placed it in the garden. For several hours the queen showed no inclination to leave her brood; at last she took wing, but though she marked the spot, she never returned.

This was disappointing, and no further attempts were made for some years—not, in fact, until 1910, when a searching *terrestris* queen having been captured and induced to lay eggs by being confined with another in the manner explained, two *terrestris* workers were caught at flowers, and after having been kept in a dark box for a few hours, were placed with her. These workers soon became attached to the brood, tending it as if it was their own, with the result that the queen did not desert, and the larvæ grew, span their cocoons, and developed into workers. In this way, therefore, a colony was successfully established.

In the spring of 1911 this experiment was re-peated with six pairs of queens, and every case proved successful. In some instances, *lucorum* workers were supplied instead of *terrestris* workers.

I also succeeded in getting several queens to start laying with only one worker as a companion, and so the duelling and killing of queens was avoided. But it was difficult to get the brood reared properly without adding one or two more workers.

In some cases I allowed the workers to fly after they had settled down, but occasionally they got

lost, and as it was early in the season and workers
were scarce, I found it best to keep them confined
until they were reinforced by the young workers
that emerged, feeding them regularly with honey
and pollen. Letting them out, however, on fine
days always had a good effect, and when at
last they were numerous enough to be allowed
continuous liberty, the brood flourished and multiplied rapidly.

The chief difficulty, indeed, in starting these
colonies was in finding sufficient *terrestris* and
lucorum workers for them. These workers do
not appear until June, and are not plentiful until
July. Fortunately *terrestris* queens, unlike those
of any other species, continue to appear for five or
six weeks after the majority of the nests have been
started, and may be found in numbers searching
for nests throughout June and in the beginning
of July.

With the object of overcoming the difficulty in
obtaining workers, I tried to get a *terrestris* colony
started with the workers of *B. pratorum*, which
appear earlier than those of any other species, and
may always be taken in plenty on the flowers of the
white dead-nettle towards the end of May.

Two searching *terrestris* queens were caught
and confined together on May 12. On May 22 I
introduced a *pratorum* worker into the nest. On
May 23 one of the queens killed the other, and
laid some eggs. The *pratorum* worker took no
notice of the eggs. On May 24 I gave two more

pratorum workers. On May 26 the queen laid more eggs, and the *pratorum* workers recognised the nest, but showed dissatisfaction with the comb, and sat apart by themselves. On June 2 the queen still sat on her brood, but the workers continued to ignore the brood, and near it had constructed some waxen cells upon which they sat. On June 10 the workers began to pay attention to the lump of brood, to which, it turned out afterwards, they had about this time added some eggs, but they continued to keep busy with their abortive waxen cells, which now numbered eight, and were of all sizes, the smallest no larger than egg-cells and built on the rims of the largest ones. On June 21 two abnormally small *terrestris* workers emerged; they had evidently been half-starved in the larval stage, the cocoons having been spun many days late. On June 24 two more dwarfed workers emerged. On July 3 a *pratorum* male was seen in the nest. On July 13 the nest contained eleven undersized *terrestris* workers, two of the original *pratorum* workers (one of these had died or been killed), three *pratorum* males, and the *terrestris* queen. The brood was not in a flourishing condition, for it consisted only of two lots of small larvæ, but this was no doubt in great measure due to the fact that the bees had been kept in confinement the whole time, a period of over seven weeks.

Evidently the company of the *pratorum* workers in the early stages of this nest encouraged the queen to rear her brood, although they were of a strange species, and gave her no assistance.

But the slow and laborious process of getting queens to start breeding in confinement, with the assistance of only two or three workers, greatly handicaps them, and it is improbable that queens so treated can ever become the mothers of very populous colonies. To give a queen caught in the fields the best possible start, she should be shut up and fed until she becomes broody, and then introduced into a de-queened nest of her own species, in which the first workers have only recently emerged. It does not do to wait until many workers have emerged, for then there is great risk of the queen getting killed, even after friendship has been established : by neglecting this particular I have lost several valuable queens.

What proved to be my most populous colony of *lapidarius* during the season of 1910 resulted from putting two small clusters of cocoons and seven large young workers with a searching queen. The large family of workers that this queen produced was chiefly due to the fact that she was able to proceed at once, while she still possessed the energy of youth, to continuous egg-laying in a prosperous colony without the delays, and, more especially, the exhausting labour, of gathering food, that are the natural experiences of a queen.

It ought not to be difficult to introduce a queen into the commencing nest of a strange species. We have seen how in nature *Psithyrus* queens are allowed to enter and dwell in weak nests, not only of the species upon which they naturally prey, but of others. On June 22, 1910, I found a queen of *B. ruderatus*, that had lost the tip of one of her antennæ, in a weak colony of *B. pratorum* that I had placed in a domicile under a wooden cover in my garden. The *pratorum* queen was also in the nest, but she seemed feeble, and died the next day. Unfortunately the *ruderatus* queen disappeared.

But whether a *Bombus* queen could be got to lay eggs in the nest of an unallied species I am at present unable to say. If this were possible one could easily rear colonies of rare species, and study their habits from queens caught in the fields.

PLACING QUEENS IN EMPTY NESTS

The method of endeavouring to get a colony of humble-bees started that would occur to most people would be to place a queen in an empty nest in the hope that she would adopt it. Many times have I tried this, always (excepting on the occasions mentioned below) with the disappointing result that the queens flew away and never returned. When I was a boy I used often to catch a queen and put her into a box containing a nest. Sometimes I fed her and shut her up for a day or two before letting her out, and then I always noticed that she circled around the spot marking it as she flew away, which seemed to show that she had some

idea of returning. In the middle of June 1907,
when driving along a road near Sandwich, my
attention was arrested by about twenty *lapidarius*
queens endeavouring to burrow into a grassy bank
facing north, the area of which was only about
thirty square yards. The opinion I formed of this
strange behaviour of so many queens was that they
had been hibernating in the bank and were endeavouring to return to their burrows in obedience
to a strong homing instinct.

On May 23, 1911, two *ruderatus* queens that I
had kept together in captivity for four days managed
to escape through a hole in the bottom of my bee-
house. One of the queens was seen shortly afterwards endeavouring to burrow into the ground
close to this hole.

Encouraged by the return of this queen, I determined to make a careful trial of placing queens
in empty nests. On May 30 I selected three of my
terrestris queens that had been confined three or
four days, and at dusk placed them in three of my
underground domiciles, one in each domicile, after
having sketched in my note-book the shape of the
lacerations on the edges of their wings so that I
should know each of them again. I had previously
placed a bees-wax cell full of diluted honey in each
of the nests. On June 2, at 8.30 P.M., I visited
each domicile, lifting off the cover and tapping
lightly on the nest material, hardly expecting the
nests to be occupied. Two of the nests gave no
answer, but my tapping on the third produced an

angry buzz and out rushed a *terrestris* queen! Next morning at 7.0 this queen was seen at the mouth of her tunnel busily trying to bite off bits of roots and grass and dragging all the loose pieces she could find into the hole. Carefully regarding her I saw that the indentations in her wings corresponded with those that I had sketched of the queen that I had put into this nest five days previously; so there was no doubt about her identity. I dropped bits of nest material and grass roots close to her, but when she discovered these she got alarmed, and spreading her wings she flew straight away, fortunately without seeing me. At 7.30 P.M. she was again at home, and I noticed that the nest material that I had dropped had been drawn into the mouth of the tunnel, which was so much choked with it that it was rendered quite inconspicuous.

On June 17 I made an examination of the nest and found that the honey-pot I had provided had not been utilised, and that the nest had been made in quite a different part of the nest-material. There were nine cocoons, larger than usual for the first batch, with some eggs on top of them and the honey-pot close by.

On the same day, wishing to get a photograph of a queen sitting on her brood, I selected this one, and after catching her in my net I carried the nest to a suitable spot for taking the picture. To make a satisfactory exposure it was necessary for the queen to sit still for about half-a-minute, and several attempts were a failure; but a successful one was finally made, and the result is shown in the frontispiece.   During the long ordeal, which lasted two hours, the queen took wing and flew back to her domicile four times.   Each time I caught her in my net, and on the last two occasions she was quite pleased to find herself confined therein, having quickly learnt that this was the prelude to coming back to her nest, and she showed great eagerness to find her brood when she was placed on the photographing table, knowing perfectly well that it was there.   Her coat was a little dusty, and she allowed me to brush it clean with a camel's hair brush as she sat on the brood, just before her picture was taken.

This nest eventually developed into a very populous colony.

No doubt by taking various precautions, which could be ascertained by experiment, in placing queens in empty nests, the proportion of successes could be much increased.

COMPLETE DOMESTICATION

A domestic animal is one that passes its whole existence under the control of man.   By controlling the pairing of domestic animals many useful and ornamental breeds of them have been produced. The honey-bee is not completely domesticated, for mating takes place in the air at some distance from the hive, and cannot be controlled in the ordinary way.[1]   But the case of the humble-bee is somewhat different, for although it is usual for the queen to fly away from the nest to get fertilised, copulation will, under certain conditions, take place in the nest.   Indeed, the difficulty with humble-bees is not so much in getting them to pair as in hibernating them.   But, from experiments that I have made with *B. lapidarius* I feel convinced that both pairing and hibernation can be accomplished under human control.

On September 5, 1910, I placed six queens and twelve males from my *lapidarius* nests in a large box covered with wire-cloth, having four inches of loose earth in the bottom.   I gave them an artificial flower consisting of a disc of red cloth tacked to the top of a stick and surmounted with three beeswax cups fastened to the cloth with melted beeswax. These cups I filled night and morning with diluted honey through the wire-cloth by means of my bulb syringe, and the bees soon learnt to come to them. On September 8 and 9 there was much copulating. On the 9th, two holes in the earth, evidently where queens had attempted to burrow, were observed. On the 10th two more holes were made in the earth.   But none of the queens buried themselves, and as they continued in a very active and restless state I let them all fly three days later.

On August 16, 1911, the attempts were renewed, and I confined ten queens in one very large box,

---

[1] Except by colour selection.   See my papers on this subject in the *British Bee Journal*, December 1909.

and six in a smaller one, with about double these numbers of males, but instead of placing loose earth in the bottom of the boxes I removed the bottoms and stood the boxes on the ground in shady places. In each box I placed a wire-cloth tray containing moss. During the first five days I filled the artificial flowers with diluted honey, morning and afternoon, then I gradually reduced the supply, stopping it altogether on August 27.

On September 5 I made an examination of the large box containing the ten queens. None of the queens had buried themselves in the moss as I had hoped, but all (except two that died) had crept in under the wire-cloth where they had fallen into a stupor. I killed one and found that her honey-sac was full of honey as in a hibernating queen. Unfortunately all the queens woke up and flew away, but if I had not disturbed them there seemed to be no reason why they should not have wintered successfully.

On September 9 four of the queens in the small box were found to be dead. Another queen was hibernating in the moss, for there was a low feeble buzz when I touched the moss. But the next day it was seen that this queen had crept out, no doubt because she was disturbed. On September 18, at 6 p.m., a cold evening, I cautiously lifted the tray in this box, and found the remaining queen in a torpid condition underneath it. Unfortunately, on the next day, this queen was also seen crawling about. Evidently the least disturbance causes the queens to evacuate their hiding-places.

By controlling breeding in the case of *B. lapidarius*, the yellow-banded variety of this species, now so rare, could be easily established as a domestic breed, and the yellow band broadened and brightened. Also the British race could be crossed with the more prolific races found in Europe, with the expectation that a breed would be obtained that would produce a much larger number of workers, a similar crossing having had this effect in the case of the honey-bee.   Possibly with controlled breeding closely-allied species could be bred together and interesting hybrids obtained.

VIII

HOW TO DISTINGUISH THE BRITISH SPECIES

PRELIMINARY REMARKS

THE humble-bees may be roughly known from other bees by their stout build and their long, thick clothing of hair.

The only solitary bees that could be mistaken for them are their nearest allies, belonging to the genus *Anthophora*, some of which are almost as burly and hairy as humble-bees; but in that genus the hair is not so long, the cheeks are so short that the mandibles appear to hinge immediately below the eyes, and the third joint of the antennæ is funnel-shaped and at least three times as long as the fourth joint. The female of *Anthophora pilipes*, one of the commonest of our wild bees in spring, has often been mistaken for a small black humble-bee, but, besides the above differences, the dense clothing of stiff golden hairs all over the outer side of her hind tibiæ distinguishes her at once. The male of this species is clothed with yellowish-brown hairs: he may be readily known from a humble-bee by the yellow markings on the skin of his face.

Certain two-winged flies of the genera *Volucella* and *Bombylius* superficially resemble the humble-bees ; they may, however, be distinguished by the absence of the hind pair of wings.

The first thing to do in examining a humble-bee is to ascertain whether it is a queen, a worker, or a male.

Notice the overlapping plates, or segments as they are called, covering the upper side of the abdomen. In the queen and worker there are six of these, but in the male seven are exposed to view. Concealed within the tip of the abdomen of the queen and worker is the sting : instead of the sting the male has a hard roundish organ about the size of a mustard seed called the armature.

The antennæ of the queen and worker have only twelve joints, while those of the male have thirteen, but this is a difference that is not so quickly seen. The antennæ of the male are longer than those of the worker and in some species than those of the queen. In all humble-bees the first joint of the antennæ, called the scape, is much longer than any of the others, which collectively are called the flagellum.

The queens are larger and stouter than the males and workers. In some species the male has yellow bands which are faint or absent in the queen and workers, but the worker is, as a rule, coloured exactly like the queen. The queens are mostly seen in

spring while the workers and the males do not
begin to appear until the approach of summer.

The skin of the humble-bee is black, so that all

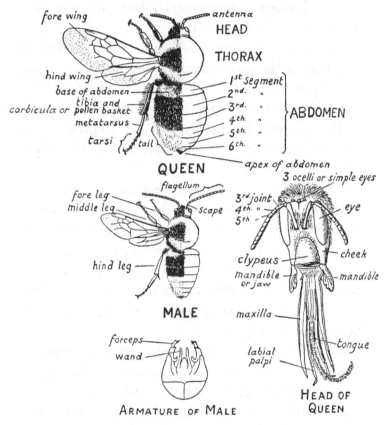

FIG. 25.—External anatomy of *Bombus ruderatus.*

colour is in the coat, but in some of the species the
feet are more or less testaceous.

The colour patterns of the coat on the upper
surface of the body differ more or less in each
species, and generally afford the readiest means of
distinguishing one species from another.   But in

L

some species they are liable to great variation, and more confidence is to be put in differences in structure, which are perfectly trustworthy when sufficiently great, although, unfortunately, they are not very numerous nor conspicuous.

Some of the species differ from one another in the length, in proportion to width of the head, more particularly of the cheeks. In the males important differences are often shown in the length of the antennæ, in the comparative lengths of the 3rd, 4th, and 5th joints of these, and in the shape of the hind metatarsi.

Many of the most closely related species differ in the quality of their coat ; thus the hairs may be short or long, equal or unequal in length, dense or thin, fine or coarse, erect or decumbent. In some cases these differences are so slight as to be only noticeable when combined with one another. When the hairs are unequal in length the coat is said to be uneven or shaggy.

One can always tell to what species or group of species a male belongs by examining through a lens the shape of the forceps and wands of the genital armature. An outline drawing of the armature of each of the British species is given on Plate VI. In comparing specimens with these drawings it must be borne in mind that the slightest change in the position of the armature, or of its parts, alters its outline, bringing fresh parts into view and causing parts previously seen to disappear, and also that the markings seen on the upper surface of the armature much depend on the way that the light falls on it, and are of no particular importance.

Variation in coat colour in a species consists chiefly in an increase or reduction of black, but in a few species the light colours may, to a slight extent, encroach upon or displace one another. On the Continent dark and light varieties, often separated geographically, of almost every species, are to be found; but in this country, except in the cases of *Bombus ruderatus* and *Psithyrus campestris*, which are often black all over as the result of dimorphism, the range of variation in blackness is, as a rule, small, and some species scarcely show any at all.

Examination of series of specimens of the different species shows that the darkening generally extends by black areas encroaching upon the yellow, red, or white ones, but there is often more or less mingling of the black hairs with the lighter ones, the latter, at the same time, losing their brightness : in some few cases, however, the pale hairs gradually deepen without any admixture of black hairs. It depends upon the species to what extent and in what parts of the body any of these processes takes place. In the queens of *B. terrestris* it is the yellow band on the thorax that is most likely to be reduced or effaced, while in *B. pratorum* it is the yellow band on the abdomen. In *B. helferanus* the thorax darkens by a mingling of black hairs and the abdomen by a deepening of yellow and brown. Yellow bands usually divide in the middle, or grow narrow, as they darken ; in many species they do both.

In the descriptions of the species that are to follow, it has been found convenient to speak of

"light specimens" and "dark specimens" of many
of the species, but these expressions must not be
taken too literally, for individuals are sometimes
found in which the degree of melanism is greater
in one part of the body than in another part.

The difficulty in naming specimens from the
colour-patterns of their coats is increased by the
fact that certain species resemble one another very
closely. But careful inspection generally reveals at
least slight differences in the extent and tint of the
colours, and in their manner of encroaching upon
one another.[1]

Another source of perplexity to the beginner
arises from the liability of the colours to fade and
bleach from long exposure to sunlight, bright red
changing to rust-red, and orange, yellow, and white
to pale brown or dingy white: black even loses

[1] An important factor in the variation in the coat colour of the humble-bees
is regional convergence, namely, the tendency of different species to show a
parallel variation in the same region. Thus Britain is within a large region
of melanism, extending into the Alps but not into the Pyrenees, which shows
itself most strongly in Schleswig-Holstein, Denmark, and Southern Scandinavia,
and is more marked in the south of England than in Scotland and Ireland.
Another kind of regional convergence is seen in *B. muscorum*, which, in Shetland
and Norway, changes to fulvous above and black underneath in sympathy
with a similar colouring of two other species, *agrorum* and *helferanus*, in
Norway. A remarkable instance of convergence is found in the Caucasian
Mountains where the yellow bands of most of the species become white. In
connection with these cases it is interesting to note that most of the North
American *Bombi* conform to a particular colour scheme, namely, greenish-
yellow, with a black tail and usually a black band on the thorax.

A further feature in the variation of humble-bees, not very noticeable within
the limits of the United Kingdom, is that the queens as a rule show greater
variation than the males; in certain regions the queens of certain species
acquire a peculiar coloration while their males show little or no change from
the type.

See Dr. O. Vogt's recent work on variation in European and Western Asiatic
humble-bees, *Studien über das Artproblem.—Über das Variieren der Hummeln*,
Berlin, Part I. 1909, and Part II. 1911.

some of its intensity and ultimately becomes shabby brown. Such fading must be allowed for in naming specimens from the descriptions given. Much exposed specimens are generally weather-beaten, and then they may be recognised by the lacerated tips of their wings and their shaggy, partly-rubbed-off coats. Occasionally, however, specimens are met with in which the colours have failed to attain their full brightness owing to their having been chilled and starved before leaving the nest.

Abnormal specimens having their coats tinged with brown (due to prolonged chill in the pupal stage), or showing irregular patches of white (due to injury during the larval or pupal stages), or with very short coats, are occasionally found. Dwarfed specimens of each of the sexes are not rare, and giant males of the *Psithyri* are now and then met with : the collector must, therefore, not be surprised if he occasionally takes a specimen that does not conform to the dimensions given.

Among the humble-bees there are several instances of two forms so closely related to one another that many entomologists have regarded them merely as races of the same species. In the British fauna there are four such couples, *B. terrestris* and *lucorum*, *B. ruderatus* and *hortorum*, *B. latreillellus* and *distinguendus*, and *Ps. vestalis* and *distinctus*. In each case the first-mentioned form is more plentiful in the south and less so in the north than its fellow. The southern form has a shorter coat, is darker, lives in larger communities, and, in three of the cases, is of a larger size than the northern form.

The relationship between *B. terrestris* and *B. lucorum*, and that between *B. ruderatus* and *B. hortorum* are closer than in the cases of the other two couples; but although these four forms are abundant over the greater part of the kingdom, living side by side, *terrestris* rarely if ever interbreeds with *lucorum*, and *ruderatus* with *hortorum*, for intermediate specimens are seldom found, and I have never known both forms to be produced by the same queen. In each case there is evidently a barrier between the two forms that does not exist; for instance, between the English and Italian races of the honey-bee which, when brought together, cross freely, and it seems best to regard them as distinct but closely-allied species.

Between a slight differentiation, such as that which separates *B. terrestris* from *B. lucorum*, and the deep divisions that part well-marked species, all degrees of relationship occur. *B. latreillellus* and *B. distinguendus* are almost as closely related to one another as *terrestris* and *lucorum*; the differences between *B. pratorum*, *B. jonellus*, and *B. lapponicus* are slightly deeper, and those between *B. derhamellus* and *B. sylvarum* deeper still: in the last case decided differences in the shape of the armature appear for the first time.

In Shetland, large-sized varieties of *B. jonellus* and *B. muscorum*, having the coat longer and more deeply coloured than it is in the mainland forms, occur. These cannot be regarded as anything more than geographical varieties, because intermediate examples are to be found in Orkney, although speculation as to their being species in the making is permissible and interesting.

All the British species of *Bombus* are found in Central Europe and ten species besides, six of which are alpine. All the British species of *Psithyrus* also occur there and one besides.

The last revision of the British *Bombi* and *Psithyri* was made by Saunders in his *Hymenoptera Aculeata of the British Islands*, published in 1896. The only alterations in the British list of species I am here making is the reinstatement as species of the three forms, *B. lucorum*, *B. ruderatus*, and *B. distinguendus*, previously recognised as distinct by Smith, the addition of *Psithyrus distinctus*, and the changing of the names *B. venustus* and *B. smithianus* to *helferanus* and *muscorum* respectively, as the result of the clearing up of the confusion between these two species.

For each of the species of *Bombus* suitable English names have been chosen in the hope that they will be of assistance to young students.

TABLE OF THE BRITISH SPECIES

Genus BOMBUS

POLLEN-STORERS

1.	B.	lapidarius	.	.	.	Stone Humble-bee.
2.	B.	terrestris	.	.	.	Large Earth Humble-bee.
3.	B.	lucorum	.	.	.	Small Earth Humble-bee.
4.	B.	soroënsis	.	.	.	Ilfracombe Humble-bee.
5.	B.	pratorum	.	.	.	Early-nesting Humble-bee.
6.	B.	jonellus	.	.	.	Heath Humble-bee.
7.	B.	lapponicus	.	.	.	Mountain Humble-bee.
8.	B.	cullumanus	.	.	.	Cullum's Humble-bee.

POCKET-MAKERS

Pollen-Primers

9.	B.	ruderatus	.	.	.	Large Garden Humble-bee.
10.	B.	hortorum	.	.	.	Small Garden Humble-bee.
11.	B.	latreillellus	.	.	.	Short-haired Humble-bee.
12.	B.	distinguendus	.	.	.	Great Yellow Humble-bee.

Carder-Bees

13.	B.	derhamellus	.	.	.	Red-shanked Carder-bee.
14.	B.	sylvarum	.	.	.	Shrill Carder-bee.
15.	B.	agrorum	.	.	.	Common Carder-bee.
16.	B.	helferanus	.	.	.	Brown-branded Carder-bee.
17.	B.	muscorum	.	.	.	Large Carder-bee.

DOUBTFUL BRITISH SPECIES

B.	pomorum	.	.	Apple Humble-bee.

Genus PSITHYRUS

1.	Ps.	rupestris	.	.		breeds in nest of B. lapidarius.
2.	Ps.	vestalis	.	.	.	„ B. terrestris.
3.	Ps.	distinctus	.	.	.	„ B. lucorum.
4.	Ps.	barbutellus	.	.	.	„ B. hortorum.
5.	Ps.	campestris	.	.	.	„ B. agrorum.
6.	Ps.	quadricolor	.	.	.	„ B. pratorum.

Note.—Closely allied species are bracketed together.

GENUS BOMBUS, Latreille.

I have divided the British *Bombi* into two main sections, the **Pollen-storers**, consisting of eight species[1] which store the pollen in cells detached from the clusters of larvæ, and the **Pocket-makers**, consisting of nine species which place the pollen in cells attached to the clusters of larvæ; and I have subdivided the Pocket-makers into the **Pollen-primers**, comprising four large long-tongued species that lay their eggs in cells primed with pollen and dwell under the ground, and the **Carder-bees**, comprising five small species that dwell, as a rule, on the surface of the ground, lay shorter and rounder eggs than the other species, and, in the male, have the joints of the antennæ swollen behind.

For the characters distinguishing *Bombus* from *Psithyrus* see p. 203.

[1] Two of these, *B. soroënsis* and *B. cullumanus*, the nests of which I have not yet found, are included because in structure they show most affinity with the pollen-storers.

MILLIMETRES.

```
|0   /|0   2|0   3|0   4|0   5|0
```

SECTION I.—*Pollen-storers.*

1. BOMBUS LAPIDARIUS, Linnæus.
Stone Humble-bee.

QUEEN.—Large ; length 20-22 mm., expanse 37-40 mm.   Shape of abdomen somewhat elongate.

Black, with the three last segments of the abdomen bright red.   Hairs of the corbicula entirely black.

Occasionally there is a narrow greyish-yellow band on the front of the thorax, or a trace of it.

The coat is rather short, dense, and fine.

Wings clear.

WORKER.—Length 11-16 mm.

Only differs from the queen in size.

MALE.—Length 14-16 mm., expanse 27-30 mm. Abdomen rather elongate.

Head black, with the face sulphur-yellow.   **Thorax black, with a sulphur-yellow band in front** and a much narrower one behind, the latter generally reduced to a few hairs and sometimes absent.   **Abdomen black, with the four last segments bright red,** and with sometimes a little sulphur-yellow on the 1st segment.

Antennæ very short—length of flagellum 4½ mm.

Armature distinct.

This is the only common species in which the queen and worker are black with a red tail except *B. derhamellus*, which is easily distinguished by the red hairs of the corbicula, and *Psithyrus rupestris*, which may be known by its brown wings.   Moreover, the red in fresh specimens of *lapidarius* is brighter than in any other species except *B. lapponi-cus*, the mountain bee, making it one of the most handsome of our native insects.

B. lapidarius is abundant in England, but less common in Scotland, and it is not found in the north of Scotland, though of late years it has become tolerably common in Orkney. It is common in many parts of Ireland. It extends across Europe into Asia, but it is not found in the central and northern parts of Scandinavia.

On the Kentish coast, in average seasons, the queens do not begin to emerge from their winter quarters until about the middle of May, and they continue appearing until the middle of June.

The usual position of the nest is under the ground : and the tunnel is generally from 18 inches to 2 feet long, but sometimes it extends to 3 feet, and in one nest I dug up it was over 7 feet. I have never found an unsheltered nest on the ground, but a favourite place is in the walls of barns and outhouses, often at a considerable height above the ground. The name *lapidarius* was given to this species on account of its habit of nesting under large stones, a position that would doubtless be to its taste in a rocky country.

A large number of workers are produced. The comb has a particularly neat and clean appearance. The cocoons are pale yellow. The wax is of a lighter brown than that of any other species, in fact it is almost yellow ; it is produced in great quantity, and the bees work it skilfully into thin sheets and cells.

This species is one of the best and most interesting to study. Among its favourite flowers are sycamore, sainfoin, birds'-foot trefoil (*Lotus corniculatus*), and the knap-weeds (*Centaurea scabiosa* and *nigra*). It is also fond of many of the flowers frequented by honey-bees.

The variety of the queen having a greyish-yellow band on the front of the thorax is rare in Britain. Smith saw only a single specimen taken near Sandwich, Kent, and for twenty years I have never noticed it. During the cold season of 1910, however, several undersized workers from larvæ that had been starved and chilled during the early part of their existence developed a prominent band.[1] In the following year (1911) the band was to be seen faintly in a number of normal specimens. In one nest taken, the queen and nearly all her workers showed it more or less, and in five out of eight nests traces of it were to be seen in some of the workers. In July 1911 a well-developed worker, showing a very perfect band, was sent me by Mr. H. Ellison from Stromness, Orkney, together with three queens, one of which showed faint traces of a band. The variation seems to be favoured by inclement weather. When the band is very faint it can only be seen by looking at the specimen from in front and it is interrupted in the middle. Very faintly-

---

[1] In 1911 I bred several freak workers of this kind, having also a faint band on the back of the thorax.

banded specimens are probably to be found in many places. I took one at Cobham, Surrey, in 1911.

The male corresponding to the banded queen has a yellow band on the back as well as on the front of the thorax, and the 1st segment of the abdomen yellow. Queéns banded with bright yellow like this male are to be found in the Pyrenees and in Italy.

A number of impregnated *lapidarius* queens from England that I was asked to supply were set free near Christchurch, New Zealand, in 1906 and 1907, but Dr. Hilgendorf, of the Canterbury Agricultural College, informs me that up to November 1911 no specimens have been seen flying ; so the attempted introduction of this species into New Zealand is probably a failure.

Bombus mastrucatus, Gerst., is an interesting species related to *B. lapidarius*, occurring in the Alps and Pyrenees at altitudes ranging from 3000 to 7000 feet. It is very like *B. lapidarius* in colouring, but it is slightly larger, the coat is long and shaggy, and the mandibles are toothed. In the armature the forceps are shaped much as in *lapidarius*, but the wands are like those of *B. pratorum*.

2. BOMBUS TERRESTRIS, Linnæus.

Large Earth Humble-bee.

Synonym :—**virginalis** (Kirby), according to Smith.

QUEEN.—Large ; length 20-22 mm., expanse 40-42 mm.

Black, with a band of deep yellow across the front of the thorax and another **on the 2nd segment** of the abdomen, and with the **tail tawny,** the zone of this colour usually commencing on the 4th segment, near its edge.

Dark specimens have the yellow band on the thorax narrow and dusky or absent, and often the yellow band on the abdomen brownish.

WORKER.—Length 11-17 mm.

Resembles the queen, but the tail is white or tawny-white, always, however, **shading into tawny at its extreme base.**

MALE.—Length 14-16 mm., expanse 30-33 mm.

Black, with a **yellow band across the front of the thorax,** a **yellow band on the 2nd segment** of the abdomen, and the **tail tawny-white,** the zone of this colour usually commencing on the 4th segment.

Antennæ short ; length of flagellum 5 mm.

Armature almost exactly like that of *B. lucorum.*

I possess a specimen from Ripple, in which the tail is black.

In England this fine humble-bee is probably more abundant than any other species, but in Scotland and the north of Ireland it is outnumbered by *B. lucorum,* and I know of no specimens from Orkney or Shetland.

Abroad it has a wide distribution, extending over almost the whole of Europe and southwards to the north coast of Africa, where it breeds in the winter and sleeps in the summer. It has also been accli-

PLATE I.

Bombus terrestris.

Bombus lucorum.

Bombus lapidarius.

Bombus soroënsis.

ALL NATURAL SIZE.

matized in Australia and New Zealand. Over the greater part of Europe the tail is white like that of *lucorum*. In Corsica the yellow bands are absent and the tail and legs red in the queen and worker (var. *xanthopus*). In a variety found in the Canary Islands (var. *canariensis*, Pér.) the yellow bands are also absent but the tail is white.

Although the earliest queens are among the first humble-bees to appear in the spring, the majority of the nests are not started until May. The nest is almost always under the ground, a long hole being preferred. The cocoons are only loosely connected to one another, the clusters being indistinct. Not many workers are reared by the queen in the first batch, but ultimately the worker population generally exceeds two hundred. The workers are most energetic and particularly industrious in gathering pollen : they defend the nest bravely when it is disturbed, hovering around it for some time ready to sting anything that approaches. The old queen, after she has laid many eggs, loses the greater part of her hair. The wax is dark brown.

This species frequents many of the same kinds of flowers as *B. lapidarius* ; the workers are also fond of the flowers of the blackberry, the numerous stamens in which seem to irritate them, for they shake their wings and buzz impatiently as they rifle each blossom.

3. BOMBUS LUCORUM, Linnæus.

Small Earth Humble-bee.

Closely related to *B. terrestris*.

QUEEN.—**Slightly smaller than** *terrestris* ; length 19-20 mm., expanse 36-39 mm.

Resembles *terrestris*, **but the yellow bands are of a lemon tint** and the **tail is white.** The yellow bands are less inclined to melanism than in the case of *terrestris*.

WORKER.—Length 11-17 mm.

Differs from the queen only in size. **Distinguishable from** *terrestris* worker **by the pure white tail.**

MALE.—Length 14-16 mm., expanse 30-33 mm.

The colouring may be described as that of *terrestris*, with the **yellow paler and more widespread**, and with the **tail pure white** instead of tawny-white.

Hairs on the face and on the top of the head lemon-yellow. Thorax with a broad lemon-yellow band in front, and a narrower, more or less indefinite band of the same colour or of whitish behind, the space between the bands being black, more or less suffused with white or lemon. The abdomen in the majority of specimens has the 1st and 2nd segments lemon-yellow, with the 3rd and part of the 4th segments black, but in the yellowest specimens the first four segments are lemon-yellow, with only a black band on the 4th segment ; the white of the tail, which usually commences on the 4th segment, shows no tendency to tawniness.

The black hairs are often mixed with white or white-tipped hairs, and there are sometimes narrow white bands on the 3rd and 4th segments.

Dark specimens are coloured almost like *terrestris*, but the tail remains pure white.

The coat is longer and less even than in *terrestris*.

Antennæ short as in *terrestris* ; length of flagellum 5 mm.

Armature almost identical with that of *terrestris*.

B. lucorum is common over almost the entire kingdom, and is especially abundant in the north, but there is no record of it from Shetland. In the north of Scotland and the Orkney Islands the queens are as large as those of *terrestris*. This is the variety *magnus* of Vogt.

Specimens intermediate between *terrestris* and *lucorum* are occasionally met with; probably most of these are hybrids.

The queens appear in April and May, and the colonies are established and mature earlier than the majority of those of *terrestris*. The nests are almost always under the ground, generally with a shorter tunnel than those of *terrestris*. I once found a nest on the surface under a box, and another in a stack of straw. The workers number 200 and more in populous nests; they are milder tempered than those of *terrestris* and seldom attack the disturber. The comb resembles that of *terrestris*. As the queen grows old, unlike the *terrestris* queen, she loses very little of her hair.

The differentiation between *terrestris* and *lucorum* is maintained on the Continent, where *lucorum* is found more abundantly in the north and in mountains.

M

4. BOMBUS SOROËNSIS, Fabricius.

Ilfracombe Humble-bee.

QUEEN.—Small ; length 15-17 mm., expanse 28-32 mm.

British specimens are **coloured very like** *lucorum*, but they may be known by their **much smaller size** and **by the shape of the yellow band on the 2nd segment of the abdomen, which is either interrupted in the middle or extends on to the sides of the 1st segment.**

The following additional characters, though less noticeable, will help to confirm identification :—

The colour of the hairs on the 6th segment and of those fringing the last three ventral segments is largely reddish-golden ; in *lucorum* there is little or no tendency to this colour. The white of the tail is inclined to be dingy, and is sometimes tinged at the base and sides with red. As in *pratorum*, the tint of the yellow band on the thorax is slightly deeper than that of the yellow band on the abdomen. The coat is shorter and more decumbent than in *lucorum*, and somewhat shorter and more even than in *pratorum* ; it is somewhat less dense on the middle of the 1st segment, and also on the middle of the 2nd segment, at the base, than in *lucorum*. The head and cheeks are slightly narrower in proportion to their length than in *lucorum*. Last ventral segment less compressed than in *lucorum*. Inner side of auricle not hairy.

In every British specimen I have examined there is a minute testaceous spot on the corbicula near the limen.

WORKER.—Length 10-14 mm.

Resembles the queen except in size. In most British specimens the tail is entirely white, but some have it tinged with red at the base.

MALE.—Length 12-14 mm., expanse 25-28 mm.

The **colouring is rather like that of** *lucorum* or *pratorum*, but the face and top of the head are black with little or no trace of yellow. Thorax black, with a yellow band in front. Abdomen black, with the 1st and 2nd segments yellow, but usually some black at the base of the 1st segment, and with the tail colour, which commences on the 4th segment, sometimes entirely white, but often white with red at the base, or red grading into white at the tip.

The hind metatarsi are slender and taper gradually towards the base, where they are very slender. By this character the male of *soroënsis* may be easily known, the metatarsi being approximately parallel-sided up to within a short distance of the base in all the other species except *terrestris*, *lucorum*, and *cullumanus*: in these three the metatarsi are wider, and are fringed with shorter hairs than in *soroënsis*.

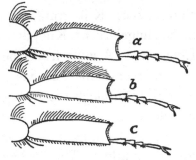

FIG. 26.—Hind metatarsus of (*a*) *terrestris* and *lucorum*, male ; (*b*) *soroënsis*, male ; (*c*) *pratorum*, male.

Antennæ with the 3rd joint shorter than the 4th ; in *terrestris* and *lucorum* the 3rd joint is longer than the 4th. Length of flagellum 5½ mm., thus slightly longer than in *terrestris*, *lucorum*, and *pratorum*.

Armature distinct.

In Britain *B. soroënsis* is local and often rare. The workers were taken abundantly at Ilfracombe on bramble blossom by the late Edward Saunders. The queens have been found in considerable numbers by Mr. E. B. Nevinson at Abersoch, on the south coast of Carnarvonshire, at the flowers of the raspberry in the beginning of June. Mr. Nevinson has also taken it at Criccieth, Bude, and in

Berkshire at Streatley.   Other localities are—
Shotover, Oxfordshire, and in Berkshire, Tubney,
near Abingdon, and Charlbury (Hamm) ; Boxmoor,
Herts (Piffard) ; Wicken, Cambs, on comfrey
(Nevinson) ; Brighton Downs (S. S. Saunders) ;
Croydon and Ipswich (Rothney) ; Barton Mills,
Suffolk (Perkins) ; Rugby (Morice) ; Yarm,
Yorkshire (Rudd) ; Carlisle (Heysham) ; in Ayr-
shire at Barr, Kilkerran, and Dundonald (Dalglish);
in Stirlingshire at Kilsyth, and in Lanarkshire at
Elvanfoot.   A male from the last-named locality in
the Saunders' collection has the tail darkened by the
admixture of black hairs, except a small spot in the
middle.

On the Continent the coloration of this species
varies greatly.   The yellow bands are sometimes
absent, and a variety in which the tail in all sexes is
red, is, in many places, commoner than the white-
tailed form.

5. BOMBUS PRATORUM, Linnæus.

Early-nesting Humble-bee.

QUEEN.—Small ; length 15-17 mm., expanse 28-32 mm.

Head and thorax black, with a **yellow band across the front of the thorax.** Abdomen black, with a **yellow band on the 2nd segment,** often interrupted in the middle, and with the **5th and 6th segments,** sometimes also the 4th, **red.** The black hairs are often tipped with white.

Dark specimens have the yellow band on the thorax more or less interrupted in the middle, and the yellow band on the abdomen much interrupted or absent.

Coat uneven. Compared with *terrestris* the difference is especially noticeable in the yellow band on the 2nd segment of the abdomen, the hairs constituting this band being unequal in length and decumbent, while in *terrestris* they are equal and erect.

WORKER.—Length 10-14 mm.

Coloured like the queen, but the yellow band on the abdomen is often absent.

MALE.—Length 11-13 mm., expanse 23-26 mm.

Hairs on the face and on the top of the head yellow. **Thorax** black, **with a broad yellow band in front** and sometimes a few yellow hairs behind. Abdomen black, with the **1st and 2nd segments yellow,** and the **5th, 6th, and 7th** segments, sometimes also the 4th, **red.** The black hairs are often tipped with white.

Dark specimens have the yellow on the 1st and 2nd segments of the abdomen obscured or absent, and only the 6th and 7th segments red.

Coat long and shaggy.

Antennæ rather short ; length of flagellum, 5 mm.

Armature like that of *B. jonellus* and *lapponicus.*

B. pratorum is a very common species in England and Scotland, but there is no record of it from Ireland. In the south of England the queens appear in March and April, and they commence nesting earlier than those of any other species; the males are plentiful in June, and the colonies break up in July. Occasionally, however, specimens may be seen in August and September. During the first week in September 1911, some *pratorum* workers were busy on the flowers of scarlet runner beans in a garden at Ripple when the workers of all the other species were lethargic and their colonies on the eve of dissolution.

This species is very unconventional in the choice of its abode. Sometimes a bird's nest is utilised, it may be in an ivy-clad tree at some height from the ground : sometimes its dwelling is at a short distance under the ground, and here a decaying tree-root is a favourite place. Often the nest is situated on the surface like that of the carder-bees, but it is generally in a wood or under a bush or tree.

The colonies are not so large as those of *terrestris* and *lapidarius*, and the workers are very meek when the nest is disturbed. The wax is dark brown. The pollen is stored in waxen cells built singly on eminences; these cells are smaller, and of a more regular shape than the pollen cells of *terrestris*.

The queens are very fond of the flowers of *Ribes sanguineum*, and the workers and males of raspberry, blackberry, *Cotoneaster*, and the woody nightshade (*Solanum dulcamara*).

6. BOMBUS JONELLUS, Kirby.

Heath Humble-bee.

Synonym :—**scrimshiranus**, Kirby.

QUEEN.—Small; length 15-17 mm., expanse 27-32 mm.

The colouring is **very like** that of *hortorum*.

Face black, top of the head more or less yellow. Thorax black, with a yellow band in front and another behind. Abdomen black with the first segment, and generally the base of the second segment yellow, and with the fourth and following segments white, rarely cream-coloured, the white often beginning on the edge of the third segment. **Hairs of the corbicula usually reddish :** in *hortorum* they are entirely, or almost entirely, black.

Head short, the length of the cheeks being not more than a third of that of the eyes. By this character this species may be easily distinguished from *hortorum* and *ruderatus*.

WORKER.—Length 10-14 mm.

Resembles the queen except in size.

MALE.—Length 12-14 mm., expanse 24-27 mm.

Coloured like the queen, but the face is yellow and the black on the abdomen often extends on to the base of the fourth segment.

Easily distinguishable from *hortorum* **by its short head,** the length of the cheeks, as in the queen, being not more than a third of that of the eyes. It is also a smaller, less elongate insect than the male of *hortorum*, and the **antennæ are decidedly shorter than in** *hortorum*, the length of the flagellum being only 5 mm.

Armature like that of *pratorum* and *lapponicus*.

B. jonellus is a local species, but commoner and more widely distributed in the north and west than in the south and east. It is rare in East Kent and

I have only once found a colony; this was in a
squirrel's nest in the top of a Scotch fir. It frequents heaths and is very partial to the flowers of
Erica.   Mr. Nevinson tells me he takes it also on
bilberry.   I have taken the males in October in
Connemara on the fuchsia that grows wild there.
Mr. Nevinson says that the hum of this species
is almost as shrill as that of *B. sylvarum*.

B. jonellus, variety nivalis, Smith.

This variety, which is found in Shetland, is larger
than the ordinary form (length of the queen 17-18 mm.,
expanse 31-33 mm.) and has the yellow bands wider
and of a deeper tint, and the tail tawny instead of white,
also the coat is longer and more shaggy.   In the queen
and worker the hairs of the corbicula are black, not
reddish.

This variety, or one approaching it, also occurs
in the Outer Hebrides, a tawny-tailed male having
been taken in Harris by Dale.

Specimens of *jonellus* that have recently been
sent me from Stromness, Orkney, resemble the
ordinary English form, except in being slightly
larger, in having a rather longer coat, and, in the
case of some of the queens and workers, having the
hairs of the corbicula black.

7. BOMBUS LAPPONICUS, Fabricius.
Mountain Humble-bee.

QUEEN.—Medium-sized ; length 16-18 mm., expanse 33-35 mm.   Abdomen short and broad.

Head black, with more or less pale yellow on top and sometimes on the face.   Thorax black with a pale yellow band in front and a much narrower one behind ; in dark specimens the latter band is absent.   Abdomen with the first segment black, or in light specimens more or less pale yellow, and with the **2nd and following segments bright red,** inclining to pale yellow at the sides of the 4th and 5th segments.   In dark specimens the black on the 1st segment spreads on to the 2nd segment, shading into the red.   Hairs on the hind tibiæ black.

On the abdomen the coat is very long, but even.

WORKER.—Length 10-15 mm.
Only differs from the queen in size.

MALE.—Length 14-15 mm., expanse 26-27 mm.
Coloured like the queen.
On the abdomen the coat is very long, but even.
Antennæ short ; length of flagellum 5 mm.
Armature like that of *B. pratorum* and *jonellus*.

This beautiful species is local and is found only in mountainous districts, with few exceptions.   It has been recorded by Smith from the Black Mountain, Brecknockshire ; Snowdon and other mountains in Wales ; Herefordshire ; Monmouthshire ; Halifax Moor, Yorkshire, and Loch Rannoch, Perthshire : and by Saunders from Gloucestershire ; Cannock Chase ; Clwyd Hill ; High Moors, Marston ; Stalybridge ; Chat Moss ; Keighley, Yorks ; Argyllshire and Aberdeenshire.   Other localities are Penmaen-

mawr (Gardner) ; Bwkhgwyn (Newstead) ; Wilsden,
Yorks (Butterfield) ; Huddersfield (Morice) ; Bude
and Malvern on bell heather (Nevinson) ; Kingussie,
Callander, Lomond Hills, Balerno, and on the Isle
of May and Bass Rock at the mouth of the Firth
of Forth (Evans) ; Golspie, Sutherlandshire (Yerbury) ; Craigellachie, Banffshire.

8. BOMBUS CULLUMANUS, Kirby.

Cullum's Humble-bee.

QUEEN and WORKER.—The queen and worker of this species have not yet been taken in Britain.

Their size is probably medium.

Specimens of large workers in the Saunders' collection from Burgos, Spain, are black with a greyish-yellow band on the front of the thorax and a somewhat narrower one at the back ; the 1st and 2nd segments of the abdomen greyish-yellow, and the **4th and following segments red.** In specimens in the same collection from Schleswig, the 4th and following segments are red, but the greyish-yellow bands are absent.

In all the specimens the corbicula hairs are black, but the longer ones are tinged with yellow. The fluff on the limen is paler than in *lapidarius.*

The coat is rather short, even and stiff. The hairs on the legs are rather shorter and stiffer than in most species. Cheeks slightly shorter than in *lapidarius*, and **more strongly punctured** at the sides and base.

I find that, under the microscope, the **hind metatarsi are not covered with a net-like pattern** as in *lapidarius*, *derhamellus*, and (the base excepted) *pratorum*. **Upper apical corner of hind tibiæ much upturned.** Auricle not hairy on the inner side as in *lapidarius*. Length of 3rd antennal joint about $1\frac{1}{2}$ times its apical width; in *lapidarius* and *derhamellus* it is about twice its apical width.

MALE.—Length 13-$14\frac{1}{2}$ mm., expanse 26-28 mm.

The colouring is very like that of an extremely light *derhamellus* male.

Face and top of the head greyish-yellow. **Thorax black, with a greyish-yellow band in front and an almost equally wide one of the same colour posteriorly.** Abdomen with the **1st and 2nd segments greyish-yellow, 3rd segment black,** the division between the black and the yellow very clearly defined, **4th and following segments red.** The underside is clothed with long greyish-yellow hairs. The hairs on the tibiæ are greyish-yellow, not red as in *derhamellus*.

The coat is denser and somewhat longer than in *derhamellus* : on the abdomen it is even, not shaggy as in *derhamellus*, and, especially on the 2nd, 3rd, and 4th segments, is more erect than in that species.

The hind metatarsi are not parallel-sided as in *derhamellus*, but narrow gradually from below the middle to the base.

Eyes rather prominent. Cheeks short, shorter than in *derhamellus*.

Antennæ with 3rd joint longer than 4th. Length of flagellum $5\frac{1}{2}$ mm. Each joint of the flagellum except the first is slightly arched, more distinctly so than in *derhamellus*, and joints 6-12 are not swelled behind as in that species.

Armature distinct.

Very rare in Britain where only the male has so far been found. The type specimen was taken by Kirby at Witnesham in Suffolk. Four more specimens were captured at Southend by Smith, who also recorded examples from Brighton Downs and Bristol. All these captures were made over thirty years ago, and none have been recorded since, but on September 8, 1911, I took a somewhat faded male at Ripple on a roadside head of *Centaurea nigra*. A colour photograph of this specimen is shown in Plate II. Probably, if it were sufficiently looked for, the species would be found existing over an extensive area in the south and east of England. It is also rare on the Continent.

This is evidently a late-appearing humble-bee, for the Southend specimens were taken at the end of August and my capture was made when the other species were beginning to disappear.

PLATE II.

Bombus pratorum.   Bombus lapponicus.   Bombus jonellus.

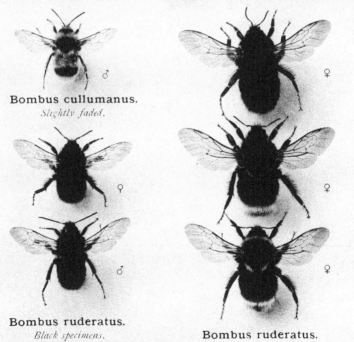

Bombus cullumanus.
Slightly faded.

Bombus ruderatus.
Black specimens.

Bombus ruderatus.

ALL NATURAL SIZE.

SECTION II.—*Pocket-makers.*

SUB-SECTION I.—*Pollen-primers.*

9. BOMBUS RUDERATUS, Fabricius.

Large Garden Humble-bee.

Synonyms :—**subterraneus** (Linn.), according to Smith ;
harrisellus, Kirby.

Closely allied to *B. hortorum.*

QUEEN.—Large ; **larger than** *hortorum* ; length 21-23 mm., expanse 40-43 mm.

The lightest examples are **coloured very like** *hortorum,* namely, with two yellow bands on the thorax, the base of the abdomen yellow, and the tail white ; but the yellow bands are of a slightly deeper and duller tint than in *hortorum,* and **the yellow band at the back of the thorax is approximately of the same width in the middle as the yellow band on the front of the thorax.** The yellow on the 1st segment of the abdomen is more or less interrupted in the middle by black, and does not spread on to the 2nd segment ; the edge of the 3rd segment is white at the sides only, and the 5th segment has often a few black hairs about the middle.

In darker specimens the abdomen has little or no trace of yellow on the 1st segment, and the white of the tail is often dingy. This variety closely resembles *B. latreillellus,* the description of which see.

Still darker specimens have the yellow bands on the thorax narrow and dusky, and the white of the tail brownish and displaced by black on the middle or whole of the 5th segment.

In the next stage of darkening almost all trace of the yellow thoracic bands has disappeared, except sometimes a pale round spot a little to the front of each wing, and the abdomen is black with the 4th segment brown.

The darkest specimens are entirely black.[1] This is the *Apis harrisella* of Kirby.

The coat is slightly more even, slightly stiffer, and slightly shorter than in *hortorum*, but markedly longer than in *latreillellus*. These coat-quality differences are most noticeable in the yellow hairs on the 1st segment when the latter are present.

The head (viewed from in front) is very long in proportion to its width, less long than in *hortorum*, but more so than in *latreillellus*, the cheeks being about half as long as the eyes.

WORKER.—Length 11-18 mm. Banded specimens and entirely black specimens occur, but intermediately coloured specimens are seldom found, unless they are very large.

The banded variety is coloured like the lightest variety of the queen, the only variation towards blackness usually met with being a slight narrowing and darkening of the yellow bands on the thorax and on the 1st segment of the abdomen ; the latter band seldom disappears, and in all examples the tail remains white.

The banded worker may be known from the workers of *hortorum* and *latreillellus* by the characters given in the description of the queen, but separation is often difficult in the case of small and worn specimens.

MALE.—Scarcely larger than *hortorum* ; length 15-16 mm., expanse 30-32 mm. Abdomen rather elongate.

Banded specimens and entirely black specimens occur,[1] but intermediately coloured specimens are rare.

Banded specimens are coloured like the male of *hortorum*, but the sides of the thorax, under the wings, are more or less yellow, while in *hortorum* they are usually black. The yellow bands are more sharply separated from the black than in *hortorum* ; those on the thorax are broad, the posterior band being as broad in the middle as the band at the front ; the yellow on the base of the abdomen is usually confined to the 1st segment, but occasionally

---

[1] The only other British species that varies to entirely black is *Psithyrus ampestris*, p. 216.

there are a few yellow hairs on the extreme base of the 2nd segment.

Intermediately coloured specimens are black, with the tail greyish-brown.

The beards on the mandibles are rust-coloured, though often discoloured with nectar; in *hortorum* they are black.

The coat is shorter, somewhat stiffer, and much more even than in *hortorum* : these combined differences are very noticeable.

Face long as in *hortorum*, but the eyes are slightly larger and the cheeks slightly shorter, these being about half as long as the eyes.

Antennæ long, as in *hortorum*; length of flagellum $6\frac{1}{2}$ mm. Armature like that of *hortorum*.

This is a very common humble-bee in England, but it is less abundant in Scotland, disappearing completely in the north of Scotland; and I know of only one specimen from Ireland, a queen taken by Mr. Freke at Borris, Co. Carlow.   The nest is under the ground, often with a long hole; but I have seldom found it, although the species is usually plentiful at Ripple.   The wax is dark reddish brown, and is very soft and sticky; the sheets and cells made of it are thick.

The division between *B. ruderatus* and *hortorum* is not very deep, and intermediate specimens, especially of the male, are sometimes found.   But in the neighbourhood of Dover I have not yet taken an intermediate queen, and in none of the nests taken by me have the two species been mixed The distinction between the two is maintained in Central Europe, but it is not recognised by Pérez in the south of France.

Most of the little differences that separate *ruderatus* from *hortorum*—namely, the larger size of the queens, the deeper tint of the yellow bands, the shorter, less shaggy coat, and the greater tendency to disappear in the north—are similar to those that divide *terrestris* from *lucorum*. The parallelism extends even to the habits, for the queens of *ruderatus* appear in their greatest numbers later in the spring than those of *hortorum*, also they are more prolific, and the colonies are more populous and break up later. *Ruderatus*, like *terrestris*, flourishes under a great variety of conditions; its distribution extends to the north coast of Africa, and even to Madeira, and it was successfully introduced with *terrestris* into New Zealand in 1885.

The black variety is not a different race from the banded variety, but both are often produced by the same parent.

A colony taken by Tuck at Bury St. Edmunds on September 10, 1898, out of a mole's nest "quite five feet away from the entrance-hole," was examined by Saunders and the result recorded in the *Entomologist's Monthly Magazine* for November 1898. It contained 8 queens of the variety having obscure yellow bands on the front and back of the thorax, and the 4th segment of the abdomen dusky white; 3 black queens, 19 banded workers, 14 black workers, 1 banded male, and 1 black male.

A populous colony that I took at Ripple on July 7, 1911, contained 51 banded workers and

49 black workers, and the old queen was of the obscurely-banded variety. In June 1911 I took a nest in which the old queen was entirely black, and kept it under observation. All the workers produced were of the banded variety, and all the first males produced were black; but later some males of the yellow variety appeared—these were probably the young of the workers. On July 17, 1911, I took a nest in which the old queen was of the lightest variety; it contained 33 workers and 3 males, all banded.

In a strong nest found by Mr. Hamm near Dorchester, Oxon, in 1911, all the individuals were black.

It is probable that the proportion of banded to black individuals produced by each queen follows Mendel's Law. It is very remarkable that the workers and males should be sharply dimorphic, while the queens show every degree of variation. The black variety is said not to occur on the Continent.

In the north of Italy and eastwards into Styria a remarkable variety of *ruderatus*, named *argillaceus*, exists. In the queen of this variety the yellow bands on the front and back of the thorax are very wide, the abdomen is entirely black, and the wings are dark brown; but the workers and males are coloured like the ordinary banded variety. Hoffer says that *argillaceus*, with *terrestris*, has the largest number of individuals in the nests. A beautiful nest brought to him by his brother in the year 1881

N

contained "almost always 300 to 400 individuals ;
but at the time of greatest prosperity far over 400.
They flew like bees in and out ; frequently they
thronged into the flight-hole exactly like honey-
bees, especially before rain, when the first drops
fell. They possessed an extraordinarily large
amount of honey ; frequently more than 100 cocoons
and honey-pots were filled with it."

In Corsica the queen of *ruderatus*, like that of
the Corsican *terrestris*, has no yellow bands and
has the tail red ; the worker resembles the queen,
except that the red tail is lighter, but the male
shows traces of the yellow bands and his tail is
orange or white. This is the variety *corsicus*.

The males of *ruderatus* select certain trees, and,
following one another from tree to tree, hover round
the foliage at a considerable height from the ground,
whereas the males of *hortorum* fly close to the ground,
as explained on page 13.

Fully-developed *ruderatus* queens are the largest
of the British *Bombi*, although large examples of
latreillellus, *distinguendus*, and *terrestris* are almost
equal to them. The tongue of *ruderatus* is about
as long as that of *hortorum*, and, like *hortorum*,
ruderatus visits the longest-tubed flowers. Its
especial favourites are the white dead-nettle, horehound, and red clover.

10. BOMBUS HORTORUM, Linnæus.

Small Garden Humble-bee.

The banded variety of the closely-related *B. ruderatus* is very like this species; for points of difference see the description of *B. ruderatus*.

QUEEN.—Medium sized; length 17-20 mm., expanse 35-39 mm.

Head black. **Thorax black, with a bright yellow band in front and a slightly narrower one behind.** Abdomen black, with the 1st segment yellow, the yellow extending on to the extreme base of the 2nd segment, the **4th and 5th segments white,** the white generally extending on to the edge of the 3rd segment, and the 6th segment black. Corbicula hairs black, rarely tipped with red.

In dark specimens the yellow bands are somewhat narrowed, and darkened by an admixture of black hairs.

Coat rather long and rather uneven.

The head is more elongate than in any other species; it is nearly twice as long as it is broad, and **the cheeks are rather more than half as long as the eyes.** By this character *B. hortorum* is easily distinguished from *B. jonellus*, which is very similarly coloured.

WORKER.—Length 11-16 mm.
Only differs from the queen in size.

MALE.—Length 14-15 mm., expanse 29-32 mm. Abdomen rather elongate.

Head black, more or less yellow on top. **Thorax black, with a yellow band in front, and another one, somewhat narrower, behind.** Abdomen black, with the **1st segment yellow,** the yellow extending on to the extreme base of the 2nd segment in the middle; **4th and 5th segments white,** 6th black in the middle, white at the sides, 7th black.

Dark specimens have the yellow bands and white tail more or less blackened.

Coat uneven and rather long.

Head very long as in the queen, the cheeks being rather more than half as long as the eyes.

Antennæ long, length of flagellum $6\frac{1}{2}$ mm.

Armature like that of *B. ruderatus*.

A male sent me from Cargill in Perthshire is entirely black, with the exception of a few scattered hairs on the 5th and 6th segments.

B. hortorum is one of the commonest and most widely distributed of the British species, and occurs, though not very plentifully, in Orkney. The queens begin to appear and to make their nests rather early in the season. In the Dover district most of the nests are started about the middle of May, but a few queens may be seen working up to the end of June, and occasionally later. On July 17, 1911, I took a nest in which the first batch of workers had not emerged, but they began coming out two days later. The queen was rather small but very active, and showed no signs of wear or exposure, which suggested that she might have been reared the same season. The males are most abundant in July, but continue appearing in small numbers till October.

The nests are not very populous and seldom, I think, contain more than 100 workers. They are generally under the ground, with a short tunnel. I once found a queen occupying a sparrow's nest in a Virginian creeper at a height of about twenty feet from the ground on the wall of a house; the nest consisted of soft hay and feathers, and under the comb there was an addled sparrow's egg.

The cocoons are of a lighter yellow than those of *ruderatus*.

The hum of this species is not so loud as that of many others, and the queen flies very quietly to and from her nest.

Irish specimens of the queen are slightly larger than English ones and have the coat slightly coarser, with the yellow thoracic bands rather wider, especially the posterior one, this being almost as wide as the anterior one. In Irish males the yellow bands are also rather wide: a male in the Irish National Museum from Milford, Co. Donegal, has the anterior band twice as wide as the black band immediately behind it. A suitable name for the Irish variety of *hortorum* is *ivernicus*.

The tongue of the queen is almost as long as the entire insect, extending to five-eighths of an inch. It can therefore extract honey from the longest-tubed flowers, such as the woundworts, honeysuckle, nasturtium, red clover, white dead-nettle, and horehound, and it prefers these to flowers that are accessible to the shorter-tongued bees, its long tongue making it difficult for it to work on the latter.

11. BOMBUS LATREILLELLUS, Kirby.[1]
Short-haired Humble-bee.

Synonym :—**subterraneus** (Linn.), according to many continental authors.

QUEEN.—Large; length 20-22 mm., expanse 38-41 mm.

Head black. **Thorax black with a yellow band in front and a very narrow one behind,** the yellow, as in *ruderatus*, rather deep (except in young specimens), and soon becoming dull and brownish with exposure ; often the band on the front of the thorax is in the middle encroached upon from behind, or divided, by a smudge of black. Abdomen black, with the **4th and 5th segments white,** and with a fringe of yellowish or dingy white on the edge of the 3rd segment, **a narrower and fainter one on the edge of the 2nd segment** ; this being often brownish and scarcely discernible, and often a few yellowish hairs on the edge of the 1st segment: in light specimens, which are not common, the 1st segment is yellow.

Coat short — very short on the basal segments of the abdomen.

The **head is elongate,** but somewhat less so than in *ruderatus*, the **length of the cheeks** being distinctly **less than half that of the eyes.**

The general appearance is very like that of a queen of *ruderatus*, but it is always possible to distinguish the one

[1] According to the generally accepted rules of priority, *subterraneus*, Linn., an older name than *latreillellus*, Kirby, should be used for this species, and it has been extensively adopted on the Continent, but I do not think it advisable to abandon the name of *latreillellus*, by which the particular colour-variety of this species that occurs in Britain and the greater part of Western Europe is universally known. The correct appellation of our bee should thus be "*subterraneus*, var. *latreillellus*," but the introduction of the name *subterraneus* into the British list at the present time would cause much confusion, because until now it has been applied by British authors to *B. ruderatus*, British specimens of which often agree with Linnæus's later description of *subterraneus*, "hirsuta, atra, ano fusco." Some systematists hold the opinion, which is not likely to meet with universal approbation, that early names applying equally to more than one species should be dropped.

PLATE III.

Bombus ruderatus.

Bombus hortorum.

Bombus latreillellus.

Bombus distinguendus.

ALL NATURAL SIZE.

from the other by the following characters :—the slightly
shorter cheeks of *latreillellus* ; the shorter coat of *latreillellus*, most noticeable on the 1st segment of the abdomen,
especially if there are yellow hairs there ; the narrow
yellowish or brownish fringe on the edge of the 2nd segment, almost always present in *latreillellus*, but seldom in
ruderatus ; and lastly, a very conspicuous distinguishing
character which, although it is not mentioned by Smith
or Saunders, holds good in every specimen I have
examined. This is in the relative widths of the yellow
bands on the thorax. In *latreillellus* **the band at the back
of the thorax is much narrower than the band at the front,** but
in *ruderatus* the bands, measured across the middle, are
approximately of equal width.

WORKER.—Large ; length 12-18 mm.
Resembles the queen, but the yellow band on the
back of the thorax is sometimes absent, and in small
specimens there is generally no trace of yellow on the
1st segment.

MALE.—Rather large ; length 15-16 mm., expanse
28-31 mm. Abdomen rather elongate.
Pale yellow with a greenish or brownish tinge, with the
exception of the following black markings. The head
black, except on the top and in the middle of the face.
**A black band across the thorax between the wings. A black
band across the 2nd segment of the abdomen, and another,**
generally narrower, **across the 3rd segment** ; these bands are
separated by a narrow band of greenish-white on the
edge of the 2nd segment ; 6th segment black in the
middle, 7th segment black. This colouring resembles
that of *Psithyrus campestris* male, the description of which
should be consulted.

In light specimens the black bands on the 2nd and
3rd segments are reduced or absent, the 6th segment is
entirely yellow, and the 7th partly so ; in the lightest
specimens, however, there are always some black hairs on
the 2nd segment, chiefly at the sides, although they are

sometimes so few that it is necessary to use a lens to see them.

In dark specimens the pale band separating the black bands on the 2nd and 3rd segments is faint and broken, and there are traces of black bands on the 4th and 5th segments.

Coat short ; hairs on the legs short.

Antennæ long ; length of flagellum 6 mm.

Armature very like that of *B. distinguendus*.

A dark male from Hayling Island in the Saunders' collection is black with the exception of a pale yellow band on the front of the thorax, an indistinct one on the back of the thorax, whitish fringes on the edges of the 4th and 5th segments, and faint whitish fringes on the edges of the 2nd and 3rd segments.

In most years, *B. latreillellus* is abundant in the Deal and Dover district; it is also common in Suffolk and in many localities in the south and east of England. In the north of England it appears to be scarce, and I can find no evidence that it has been taken either in Scotland or in Ireland.

The queens, like those of *lapidarius*, appear late, and in the Dover district may be seen searching for nests at the end of May and early in June.

The queen often rears a very large brood. Two nests that I examined in the stage just before the first workers emerged had a number of additional cocoons fastened to the rear of the main cluster. In one of these nests, shown in Fig. 27, the additional cocoons numbered nine, and those in the main cluster sixteen, making twenty-five in all, many more than the queen could spread her body over. The main cluster was easily distinguished by the groove in its centre, and by the darker appearance of the cocoons, which contained pupæ in an advanced stage, while the additional cocoons contained larvæ and young pupæ.

The population, however, does not grow so large as that of nests of *lapidarius* or *terrestris*, the males and queens being soon produced. This species,

FIG. 27.—Queen of *B. latreillellus* incubating her immense lump of brood. Notice that it has become joined to the honey-pot.

therefore, does well in a short season. The workers are unusually large.

This species has little skill in wax-building; the honey-pot is clumsily constructed, the larvæ are often imperfectly covered, and blotches of wax are allowed to remain on the cocoons.

The bees, comb and nest, have a characteristic disagreeable odour.

B. latreillellus has rather a long tongue, and like *ruderatus* and *hortorum*, it chiefly visits long-tubed flowers. I have seen it most frequently on honeysuckle and *Trifolium pratense*.

12. BOMBUS DISTINGUENDUS, Morawitz.

Great Yellow Humble-bee.

Synonym :—**elegans**, Seidl (in part).

Related to *B. latreillellus*.

QUEEN.—Large ; length 20-23 mm., expanse 39-42 mm.

Dusky yellow or greenish-yellow, soon fading to dingy brownish-yellow, with an admixture of black hairs on the head, and **with a black or dark grey band across the middle of the thorax.** The yellow is somewhat deeper on the 2nd segment of the abdomen than on the rest of the abdomen.

In light specimens the black band on the thorax is often more or less tinged with yellow at the sides.

Coat longer and denser than in *latreillellus*, especially on the basal segments of the abdomen.

Head slightly shorter than in *latreillellus*.

WORKER.—Length 11-18 mm.

Resembles the queen except in size.

MALE.—Length 15-16 mm., expanse 28-31 mm. Rather elongate.

Coloured like the queen. Closely resembles light specimens of *latreillellus*, but there is **no trace of black hairs on the 2nd segment.**

The coat is rather denser and longer than in *latreillellus* male.

Antennæ as in *latreillellus* ; length of flagellum 6 mm.

Armature almost identical with that of *latreillellus*.

This humble-bee is rare in the south of England; it was formerly taken occasionally in the London district, and sporadic examples have been recorded from Oxford, Bristol, Lyme Regis, Newton Abbott, and Truro. In the Dover district I have only taken

a single worker.   In the east of England it has been
met with at Great Yarmouth, Lowestoft, Southwold,
and Brandon.   In the midlands, and in Yorkshire,
Lancashire, and farther north it is more common,
though it does not appear to be anywhere plentiful.
I have received specimens from Perthshire and
Orkney.   It has also been recorded from the Isle
of Man ; from North Uist in the Hebrides ; and in
Ireland from Co. Kerry, Co. Wexford, Co. Wicklow,
Co. Carlow, and Co. Kildare.

Smith states that he found a nest in Yorkshire,
and " on disturbing it the bees emitted a powerful
aromatic odour ; the community, although it contained males and females, was small.[1]

[1] *Catalogue of the British Bees in the Collection of the British Museum,*
by F. Smith, 2nd edition (1876) 1891, p. 202.

Sub-section II.—*Carder-bees.*

13. BOMBUS DERHAMELLUS, Kirby.
Red-shanked Carder-bee.

Synonym :—**raiellus**, Kirby.

QUEEN.—Small ; length 16-18 mm., expanse 29-32 mm.

Black, with the **three last segments of the abdomen red.** The red is not so bright as in *lapidarius*, and soon fades to rust-yellow. **Hairs of the corbicula red.**

Coat rather thin and uneven, not long.

WORKER.—Length 10-16 mm.

Resembles the queen except in size, but sometimes the hairs of the corbicula are red only at the tips.

MALE.—Length 13-14 mm., expanse 23-26 mm.

Head black, more or less tinged with grey. **Thorax black, generally with a dusky yellowish-grey band in front and another behind. Abdomen black, with the 4th and following segments red, and generally with the 1st and 2nd segments more or less dusky yellowish-grey.** The red is less bright than in *lapidarius.* Cf. the male of *Psithyrus rupestris.*

In some specimens the 2nd segment is bright yellow or brownish-yellow. Sometimes the red spreads on to the 3rd segment.

Coat rather thin and uneven, not long.

Antennæ rather long ; length of flagellum $5\frac{3}{4}$ mm.

Each of the joints of the antennæ, from the 6th to the 12th, slightly swelled behind.

The antennæ differ from those of *sylvarum* in having the 3rd joint much longer than the 4th.

Armature distinct, but approaching that of *sylvarum.*

B. derhamellus is a widely-distributed species in Great Britain and Ireland, and common in many places, especially in the north of England. I have taken many nests, always on the surface of the ground, and frequently in long grass on roadside banks. It commences to nest earlier than any of the other carder-bees, even than *agrorum*. The cocoons are deep yellow, and the wax is very dark. In 1911 I found three nests in a suffi-ciently early stage to ascertain the number and position of the cocoons in the first batch. In each nest

FIG. 28.

Diagram of co-coons in a com-mencing nest of *B. derhamellus* (hori-zontal section).

they numbered eight, and they were arranged symmetrically in exactly the same manner —two in the middle, forming the bottom of the queen's seat, and three on either side, at a higher level, forming its sides, as shown in the diagram (Fig. 28).

The queens are very fond of white dead-nettle; this is one of the few species I have seen visiting the flowers of the ground ivy, *Nepeta glechoma*.

14. BOMBUS SYLVARUM, Linnæus.
Shrill Carder-bee.

QUEEN.—Small ; length 16-18 mm., expanse 29-32 mm.

The prevailing colour is greenish-white, often with a yellowish tinge.   On the top of the head and in the centre of the thorax this colour shades into black.   On the 2nd segment of the abdomen there are traces of a black band, at least at the sides ; the 3rd segment is black with a narrow band of greenish-white at its edge ; 4th and following segments orange, each with a narrow band of greenish-white at its edge.

The pale colour on the 2nd segment often shades into a deeper, more distinctly yellow tint, and there are often black hairs at the base of the 4th segment.

Coat rather thin and uneven, not long.

WORKER.—Length 10-15 mm.
Resembles the queen except in size.

MALE.—Length 13-14 mm., expanse 23-26 mm.
Coloured like the queen, but the red on the 4th segment is generally displaced by black.

Quality of coat as in *B. derhamellus.*

Antennæ as in *derhamellus*, but the 3rd joint very little longer than the 4th.

Armature distinct, but approaching that of *B. derhamellus.*

This species is widely distributed in England and Ireland, and common in a good many places, but it appears to be rare in Scotland, and has not been found in the north of Scotland.   The queens and workers are very agile, both in the nest and on the wing, and their hum is more shrill than that of any other species.   The nest is usually on the surface of the ground, but I have often found it under the ground with a short tunnel. The queens appear rather late in the spring, and the colonies are generally among the last to break up in the autumn, the workers being particularly industrious on the August flowers. The wax is almost as light-coloured as that of *lapidarius*; it is produced rather sparingly, and is worked into thin sheets. The nest has as clean and tidy an appearance as that of *lapidarius*. On the Continent there is a variety of *sylvarum*, named *nigrescens*, that is coloured exceedingly like *derhamellus*.

Among the favourite flowers of *B. sylvarum* may be mentioned red bartsia (*B. odontites*) and *Centaurea nigra*.

PLATE IV.

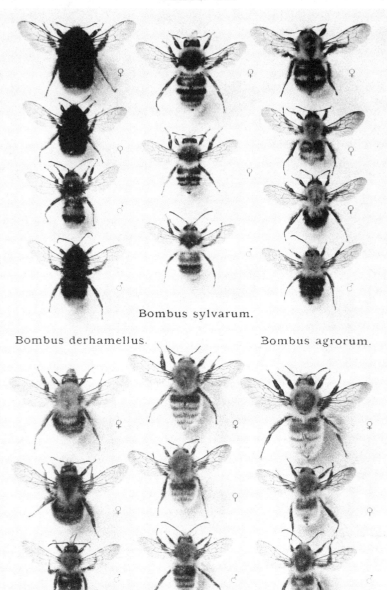

Bombus sylvarum.

Bombus derhamellus.   Bombus agrorum.

Bombus agrorum.   Bombus helferanus.   Bombus muscorum.

ALL NATURAL SIZE.

15. BOMBUS AGRORUM, Fabricius.
Common Carder-bee.

Synonym :—**muscorum** (Linn.), according to Smith.

QUEEN.—Small ; length 16-18 mm., expanse 29-32 mm.

Head clothed with a mixture of pale and black hairs.

Thorax bright tawny-yellow, often somewhat paler : black hairs often appear among the tawny-yellow ; when these black hairs are scarce they are scattered **across the front** of the thorax, when numerous they form a sooty smudge, shaped somewhat like a triangle, with its base on the front of the thorax, the back part of the thorax at the sides being left more or less unblackened.

On the abdomen the general scheme of colour is—**the base pale yellow, the middle black,** and **the tail tawny ;** but the colours frequently encroach upon one another, producing a great variety of patterns. In a common variety the black extends up and down the sides of the abdomen ; in some cases the tawny colour spreads at the same time on to the middle of the abdomen, producing a unique pattern. In another common variety the whole abdomen is black, with the edges of the segments more or less fringed with pale yellow, the fringes being widest on the apical segments ; this pale yellow may spread over the whole of the abdomen, causing the bee to resemble a somewhat faded *helferanus* or *muscorum*, but specimens that show no black hairs at the sides of the 3rd and 4th segments are rare.

The coat is rather thin, uneven, and moderately long.

WORKER.—Length 10-15 mm.
Only differs from the queen in size.

MALE.—Length 13-14 mm., expanse 24-27 mm.
Coloured like the queen.
Coat rather long and rather uneven.

O

Antennæ with **each joint of the flagellum much swollen in the middle behind.**   Length of flagellum $5\frac{1}{4}$ mm.
Armature distinct.

This species is very common throughout the United Kingdom, but there is no record of it from Orkney or Shetland.   Its nest is on the surface of the ground, often in grass or under ivy, and, like *pratorum*, it prefers to dwell in woods rather than in open country.   I have found nests in the thatch of a cow-lodge, in a tuft of pampas grass, in the grubbed-up stump of a tree, in a robin's nest, in an old shoe, and in a decayed and broken kettle, but never under the ground.

In most specimens from the Dover district the thorax is more or less blackened.   The remarkable arrangement of colour in which the whole upper side is tawny except the 1st segment and base of the 2nd segment (which are pale yellow), with black at the sides of each of the segments except the 6th, approaches that of the Norwegian variety *arcticus*, and has appeared in several workers found in a nest at Ripple.

In the south of France the whole upper side is orange (var. *pascuorum*), while in Zealand it is perfectly black (var. *mniorum*).

16. BOMBUS HELFERANUS, Seidl.
Brown-banded Carder-bee.

Synonyms :—**venustus,** Smith, this also includes *B. muscorum* ; **variabilis,** Schmiedeknecht ; **venustus,** Saunders.

QUEEN.—Small ; length 16-18 mm., expanse 30-32 mm.

The colouring resembles that of *B. muscorum* (the description of which should be consulted), and of light examples of *B. agrorum.*

Head clothed with very pale hairs, often mixed with black, especially on top and around the eyes.

Tint of the thorax on top somewhat richer than in bright specimens of *agrorum*, namely **orange-brown,** but this soon fades : black hairs often appear among the orange-brown ; when these black hairs are scarce—sometimes they are so few that a lens is needed to detect them—they are confined to the region just above and in front of the roots of the wings ; when plentiful they are scattered evenly all over the orange-brown, except in the centre.

Abdomen pale yellow, dingy white in immature examples, and pale brown in faded ones, with **a brown band on the 2nd segment,** a narrower and much fainter one on the 3rd segment, this being often scarcely discernible and in fresh specimens tinged with lemon, and sometimes a still narrower and fainter one on the 4th segment : last segment black. In dark specimens the colouring of the abdomen becomes somewhat darker, but black hairs never appear among the pale ones as in *agrorum.*

Coat rather shorter and less shaggy than in *agrorum.*

WORKER.—Length 10-15 mm.
Resembles the queen except in size. Very dark specimens are occasionally found.

MALE.—Length 13-14 mm., expanse 24-27 mm.
Coloured like the queen.

Coat somewhat shorter and less shaggy than in *agrorum*.

Joints of the antennæ less swollen behind than in *agrorum*; length of flagellum 5½ mm.

Armature distinct, but approaching that of *B. muscorum*.

This species has been so much confused with *B. muscorum* that the extent of its distribution in the United Kingdom is very imperfectly known. It is not rare in the Dover district, and has been taken, not uncommonly, in various localities in Suffolk, Cambridgeshire, Sussex, Oxfordshire, Hampshire, and Devonshire. It probably occurs throughout the south and east of England. I have seen a specimen from Manchester, and another from Helsby in Cheshire, but know of no other captures in the north. All the supposed Irish examples that I have seen have been *B. muscorum*.

In East Kent the queens appear later in the spring than those of any other species, generally not until June. All the nests that I have taken have been on the surface of the ground.

The very mild temper of this species is in strong contrast to the revengeful disposition of *B. muscorum*.

On the Continent the colouring of this species is very variable, a black variety being common in many places; this has not been met with in England.

17. BOMBUS MUSCORUM, Linnæus.

Large Carder-bee.

Synonyms :—**venustus**, Smith, this also includes *B. helferanus* ; **cognatus** (Stevens), according to Schmiedeknecht ; **smithianus**, White.

QUEEN.—Slightly **larger than** *B. agrorum* and *B. helferanus* ; length 18-19 mm., expanse 33-35 mm.

The colouring has hitherto been supposed to be indistinguishable from that of *B. helferanus*, but after having bred and compared many examples of each species, I notice that the following differences are constant, and it will probably be found that they are sufficient for identification in every case.

The **orange-brown on the thorax never has any black hairs mixed with it** ; in *agrorum* and *helferanus* black hairs are frequently present.

In specimens from England the **orange-brown on the thorax is more or less encroached upon from in front and from behind by very pale yellow**, this colour in immature specimens being almost white, and in faded ones pale brown ; this is the variety *sladeni* of Vogt.   But in specimens from Scotland there is little or no encroachment of pale hairs, and the orange-brown is slightly redder than in the south of England ; this is the variety *pallidus* of Evans.

The abdomen is slightly yellower than in *helferanus*. **The brown band on the 2nd segment** and the fainter one on the 3rd segment are **much lighter and more diffuse than in** *helferanus*, in fact they are usually only faintly discernible ; in perfectly fresh specimens they are often of a beautiful brownish lemon colour which is very evanescent and soon fades to pale brown.   The hairs on the 2nd segment look darker than they really are on account of their erectness.

The coat is **slightly longer, slightly more even, and slightly denser than in** *helferanus*, and denser and more even than in *agrorum*.   In specimens from the south of

England the coat is slightly shorter than in specimens from Scotland. Basal part of hind metatarsus less distinctly engraved than in *helferanus* and *agrorum*.

Irish examples are intermediate between the English and Scottish varieties.

WORKER.—Length 10-16 mm.
Only differs from the queen in size.

MALE.—Slightly larger than *B. helferanus* ; length 14-15 mm., expanse 26-28 mm.

Coloured like the queen, but there is little or no trace of bands on the abdomen, and in bright fresh specimens the abdomen is flushed with lemon ; this colour, however, soon becomes dingy, fading to pale brown.

Coat slightly longer, slightly more even, and slightly denser than in *helferanus*.

Joints of the antennæ slightly swollen behind, as in *helferanus* ; length of flagellum $5\frac{1}{2}$ mm.

Armature distinct, but approaching that of *B. helferanus* ; among other differences the forceps at their extremities are broader and blunter than in *helferanus*.

A male taken at Loo Bridge, in south-west Ireland, by Col. Yerbury, had the thorax dark chestnut-brown and the coat generally of a brownish tinge.—(*Ent. Mon. Mag.* for 1902, p. 54.)

B. muscorum is not an abundant species, but it is distributed throughout the kingdom. It is more common in Scotland and Ireland than in England. In Kent, and probably in other southern and eastern counties, it is chiefly confined to marshy districts. I have taken it in Romney Marsh on the flowers of the marsh mallow, and at Ripple I occasionally met with specimens in clover fields, most of which have probably strayed from the marshes near Deal.

In July 1911, having been informed by some

labourers in these marshes who were mowing the
hay there, that they frequently came across the nests
of a savage yellow bee which they feared to disturb
more than a wasp's nest, I asked them when they
next found a nest to let me know, and on July 21
was summoned to take two nests, both of this
species, situated only about ten yards apart in a hay
field.   So great was the men's fear of getting stung
that they did not dare to approach near enough to
show me the exact spots, and I found that, as soon
as I disturbed the nests, the workers flew round my
head in a most menacing manner ; they also had the
disagreeable trick of persisting in doing this, following me wherever I went for a minute or two.   In
this way the workers of *muscorum* behave quite
differently to those of *terrestris*, which only attack the
lower part of the body and will not follow one to
any great distance.   It was, however, quite easy to
remove both nests without getting stung by taking
care not to disturb them much until nearly all the bees
had been captured.   The nest material was dead
grass blades, and the stronger nest contained 52
workers, but not nearly all the workers had emerged.
There were two waxen honey-pots of a regular, tall,
and slender shape, attached to the outside of the
comb at some distance from one another.   The
cocoons were of the same pale-yellow tint as those
of *helferanus* ; the wax was a shade darker than that
of *helferanus*.   The nests were transferred to my
garden, but they did not flourish well there owing,
perhaps, to the hot, dry weather that prevailed, for

I have noticed this species to be commonest in damp cold seasons.

In Co. Armagh the Rev. W. F. Johnson found the nests in deep moss.

B. muscorum, variety smithianus, White.

This is a variety found in the Shetland Islands. It is larger than the mainland form : length of the queen 19-20 mm., expanse 34-36 mm., length of the male 15-16 mm. The queen and worker have the head (except a few hairs on top), the **underside of the thorax, the underside of the abdomen, and the legs clothed with black instead of pale hairs** ; also the orange-brown of the upper side of the thorax is of a somewhat richer tint than in the mainland variety, and it is not bordered with pale hairs ; and the abdomen is more distinctly lemon-yellow. The male is coloured like the queen, except that he generally has a tuft of lemon-yellow on the lower part of the face, some yellow on the top of the head, and the underside of the abdomen partly pale.

The coat is longer in this variety than in the mainland form.

Specimens with a black underside have also been taken in Lewis in the Hebrides. The variety occurs, too, in Norway. But specimens from Stromness, Orkney, show only traces of black on the face, underside, and legs, although the coat is long.

In the Saunders' Collection at the British Museum there is a specimen from St. Mary's, Scilly, of the South of England variety, having the underside black, and the legs partly so.

The Rev. F. D. Morice found *smithianus* nesting in stone walls in Shetland.

DOUBTFUL BRITISH SPECIES.

BOMBUS POMORUM, Panzer.
Apple Humble-bee.

QUEEN.—Length 17-18 mm., expanse about 35 mm. Head black. **Thorax black,** with an indistinct pale-yellow band in front and another behind (these bands may be absent). **Abdomen red,** with the **1st segment** and possibly the 2nd more or less **black, the black gradually shading into the red.** Hairs on the corbicula black, " frequently tipped with yellowish-red " (Hoffer).

Coat uneven and much shorter than in *lapponicus*.

Head long, shaped much as in *ruderatus*, the cheeks being long, but their length is less than half that of the eyes.

WORKER.—Probably differs from the queen only in size.

MALE.—Length 14-16 mm., expanse 26-27 mm. Shape rather slender like that of *latreillellus*.

Head black. **Thorax black, with a dingy yellowish-grey band in front and another behind.** Abdomen with the **1st segment yellowish-grey and the remaining segments red,** this colour being darker on the 2nd segment than on the rest of the abdomen.

Coat rather uneven and rather short, especially on the abdomen. Hairs on the legs short. The clypeus is hairless except at the sides and on top, and the **mandibles are unbearded.**

Cheeks long as in the queen. Hind tibiæ convex and covered with short hairs.

Antennæ of medium length ; length of flagellum 5 mm.

Armature distinct.

It is a question whether this species ought to be retained in the British list. Smith caught three

males in 1837[1] or in 1863,[2] which of these dates is
not clear, and his son a queen in 1864, near Deal,
but no other captures have been recorded. The
species occurs in Central Europe. In the mountains
of Switzerland and Austria, a variety having the
whole coat yellowish-grey with the exception of a
black band across the middle of the thorax is to be
found.

Hoffer took three nests of the red-tailed form on
the Geierkogel in Styria. They were all under the
ground, and in two of them he noticed what he
called "pollen cylinders." In one of the nests with
about 300 cells there were twelve of these pollen
cylinders. They always stood at the end of a group
of cells, but as the nature of these cells is not explained it is not clear whether the species is a pollen-
storer or a pocket-maker. The convex hind tibiæ
of the male and the long cheeks indicate that it
is probably a pocket-maker occupying an intermediate position between the pollen-primers and the
carder-bees.

[1] *Ent. Annual* for 1865, p. 95.
[2] *Catalogue of British Bees*, 2nd edition, p. 206.

GENUS PSITHYRUS, Lepeletier.

Synonym :—**Apathus** (Newman).

Distinction between Bombus and Psithyrus.

QUEENS. — *Psithyrus* queens may be readily known from *Bombus* queens by the absence of the pollen-collecting organs on the hind legs, **the outer side of the tibia being convex, dull, and clothed all over,**

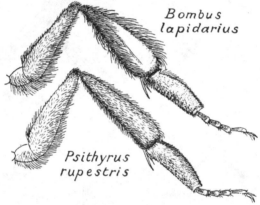

Bombus lapidarius

Psithyrus rupestris

FIG. 29.—Hind leg of *Bombus* and *Psithyrus* female.

though not thickly, with hairs, not flattened, bare, and shining in the middle as in *Bombus*; and the **comb at the apex of the tibia and the auricle at the base of the metatarsus being absent.**

In the *Psithyrus* queens **the apex of the abdomen is more or less incurved,** and close to it, **on the 6th ventral segment,** *i.e.* on the last segment on the underside, **there is a pair of elevations, the shape of which differs widely in each of the well-marked species,** and is very helpful in distinguishing them from one another.

The coat is less dense than in the *Bombus* queens, **the hair being particularly scanty on the abdomen,**

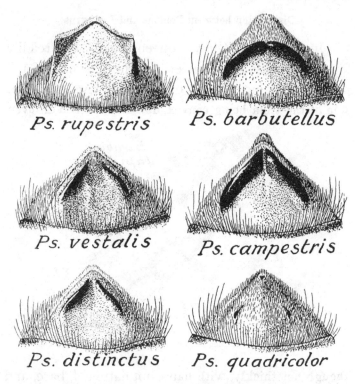

Ps. rupestris   Ps. barbutellus

Ps. vestalis   Ps. campestris

Ps. distinctus   Ps. quadricolor

FIG. 30.—Tip of the abdomen (underside) of the queen of each of the British species of *Psithyrus*, showing the elevations on the 6th ventral segment.

which in the middle looks almost bare; also the hairs are somewhat coarser than in *Bombus*.

Psithyrus queens usually have their wings more clouded than *Bombus* queens, and they may generally be recognised in the field by their weaker flight with its **distinctly lower-pitched and softer hum.**

MALES.—The beginner does not find it easy to distinguish the *Psithyrus* males from those of *Bombus*, there being no very striking characters to rely upon. At first glance they differ most noticeably in their somewhat thinner and coarser coats, and particularly in **their more shining, because less densely-clothed, abdomens.** When the 1st segment is clothed with yellow hairs the brightness of the yellow is reduced by the black skin underneath. The face is rounder and shorter than in the males of most species of *Bombus*. The forceps of the armature are pale and membranous, while in *Bombus* they are brown and horny. Some of the best distinguishing characters are to be seen in the **hind tibiæ**: these, as in the females, are, on the outer side, **transversely convex and hairy all over,** while in the males of *Bombus* the tibiæ are **flattened, bare (or almost so) in the centre, and more shining than in** *Psithyrus*, except in the males of the carder-bees, which in respect of these tibial characters occupy an intermediate position ; but as the males of the carder-bees have the joints of the antennæ swelled underneath (only the last six joints in *B. derhamellus* and *B. sylvarum*), a peculiarity not found in the *Psithyri* and in the other *Bombi*, they can easily be excluded from comparison.

1. PSITHYRUS RUPESTRIS, Fabricius.

QUEEN.—Large ; length 20-23 mm., expanse 40-45 mm. ; sometimes smaller.

Black, with the 4th and following segments of the abdomen red. The red is less bright than in *B. lapidarius.*

There is occasionally a dingy yellowish-grey band across the front of the thorax.

Wings dark brown, thus darker than in any other species.

The 6th dorsal segment is dull and is covered with very short red hairs.

The elevations on the sides of the 6th ventral segment are very prominent and form angular wings, which project so far that they can be seen plainly on either side when the tip of the abdomen is viewed from above.

MALE.—Length 15-17 mm., expanse 30-33 mm.

Head black. Thorax black, usually with indistinct **yellowish-grey bands in front and behind.** Abdomen black ; with **the sides and edges of the 1st and 2nd segment** usually more or less yellowish-grey, and with the 4th and following segments red. The red usually extends on to the 3rd segment where it shades into the black ; it is less bright than in *B. lapidarius.* In some specimens there is no yellowish-grey on the thorax and abdomen.

Coat rather long, uneven, and rather coarse.

Antennæ rather short ; length of flagellum 5 mm. This species may be easily known from the male of *B. derhamellus*, which in coloration it much resembles, by the thicker and shorter antennæ and somewhat longer coat. The antennæ are thicker than in *B. lapidarius.*

Armature distinct.

A specimen of the male sent me from Colchester is entirely black. Another taken at Ripple has the abdomen red except the 1st segment.

This large species is parasitic on *Bombus lapi-*

PLATE V.

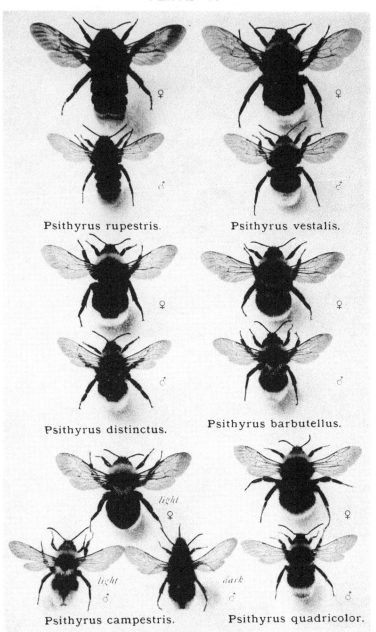

Psithyrus rupestris.   Psithyrus vestalis.

Psithyrus distinctus.   Psithyrus barbutellus.

Psithyrus campestris.   Psithyrus quadricolor.

ALL NATURAL SIZE.

PLATE V.

darius and is to be found in most places where its
host is abundant. It is plentiful in East Kent,
where it victimises from 20 to 40 per cent of the
colonies of *B. lapidarius*. It is also particularly
common in many parts of Suffolk and Norfolk. In
Ireland it has been recorded from Co. Cork, Co.
Limerick, and Co. Sligo. The variety of the female
having a pale band on the thorax is rare, but it
occurs occasionally at Ripple, and has been taken
near Norwich.

This is the latest of all the humble-bees to
awaken in the spring; it may be seen throughout
June searching for the nests of *B. lapidarius*.

2. PSITHYRUS VESTALIS, Foucrier.

QUEEN.—Large ; length 20-22 mm., expanse 40-43 mm. ; sometimes smaller.

Head black. Thorax black, with a deep yellow (often brownish-yellow) **band in front. Abdomen black ;** with the **3rd segment lemon-yellow on the sides,** except at the base, **4th segment white, 5th segment white on the sides.** The deep yellow band on the thorax is often narrow and darkened.

6th dorsal segment shining.

The elevations on the sides of the 6th ventral segment are ridge-like and rounded, and converge toward the apex but they do not nearly meet there, and the space between them there is thickly covered with short, velvety, golden hairs.

MALE.—Length 15-17 mm., expanse 30-33 mm.

Head black, often with a few yellow hairs on top. **Thorax black, with a yellow band in front** and sometimes a few yellow hairs behind. **Abdomen black ;** with sometimes the 1st segment yellow (this band is not very prominent owing to the paucity of the hairs) ; **with the 3rd segment yellow,** but black in the middle at the base ; and with **the 4th and 5th segments white,** the 6th segment more or less black in the middle and white at the sides, and the 7th more or less black.

Coat rather short ; and **differing from that of every other species of** *Psithyrus* **in being even,** not shaggy.

Antennæ long and thick ; length of flagellum 6 mm. **The 3rd joint of the antennæ is much shorter than the 5th joint, which almost equals the 3rd and 4th together.**

Armature very like that of *Ps. distinctus*, but the inner border of the forceps is densely fringed with short stiff hairs.

This species preys on *Bombus terrestris*. It is very common in the south and east of England, and probably occurs wherever *B. terrestris* is abundant. In the north, however, it is displaced by its close ally *Ps. distinctus*, with which many collectors have hitherto confused it.

The only Irish specimen I have seen is a female in the Irish National Museum, taken by Mr. Pack-Beresford at Fenagh, Co. Carlow.

3. PSITHYRUS DISTINCTUS, Pérez.

Closely related to and much resembling *Ps. vestalis*.

QUEEN.—Smaller than *Ps. vestalis*; length 18-20 mm., expanse 36-39 mm.; sometimes smaller.

The coloration resembles that of *Ps. vestalis*, but the tint of the yellow band on the front of the thorax is lighter and is best described as dull lemon; this band is always broad: sometimes there are a few pale yellow or yellow-tipped hairs behind the wings and at the back of the thorax. On the back of the thorax and on the edges of the 1st and 2nd segments of the abdomen the black shades into dark grey (brown in faded specimens) or sometimes into a paler tint. The yellow at the sides of the 3rd segment is very pale and soon fades· to white.

A giant specimen from Aviemore, in the Saunders' Collection, has a narrow yellow band on the back of the thorax and the 1st segment of the abdomen yellow on the base at its sides.

The coat is denser and, in proportion to the size of the insect, somewhat longer than in *vestalis*. The metatarsi of the hind legs are somewhat shorter than in *vestalis*: in *distinctus* the length of this joint is equal to that of all the tarsi together, but in *vestalis* it is somewhat longer than these.

The 6th dorsal segment (seen through a good lens) is very faintly and shallowly punctured; in *vestalis* it is more deeply punctured, especially at the sides.

The elevations on the sides of the 6th ventral segment are shaped very much as in *vestalis*, but they are rather higher towards the apex and the golden hairs there are shorter.

MALE.—About the same size as *vestalis*; length 15-17 mm., expanse 29-32 mm.

Coloured like *vestalis*, except that the yellow on the head thorax, and base of the abdomen is of a lighter tinge, almost sulphur-yellow in fresh specimens, and is more extensive, and the yellow on the 3rd segment is paler, soon fading to white.   In fact, the coloration is very like that of *Ps. barbutellus.*

Head black, with some yellow hairs on top.   Thorax black, with a yellow band in front and some yellow hairs, generally sufficient to form a narrow band, behind. Abdomen black ; with the 1st segment yellow ; 3rd segment light yellow, soon fading to white, but black in the middle and at the base ; 4th and 5th segments and sides of the 6th segment white.   **A variety occurs in which the 4th and 5th segments and sides of the 6th segment are yellow.**

Coat longer, much less even and somewhat denser than in

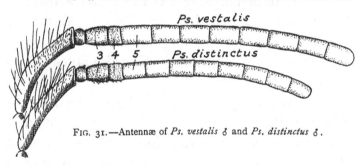

FIG. 31.—Antennæ of *Ps. vestalis* ♂ and *Ps. distinctus* ♂.

vestalis—these differences combined are very noticeable. The coat is somewhat shorter, somewhat denser, and some-what finer than in *barbutellus.*   Hairs on the legs longer than in *vestalis.*

Hind metatarsi slightly shorter than in *vestalis.*

The **best characters for distinguishing** *distinctus* from *vestalis* **lie in the male antennæ.**   These are **much shorter than in** *vestalis*—**length of flagellum 5 mm.**—and have the **3rd joint about as long as the 5th, the 5th being much shorter than the 3rd and 4th together.**

Armature like that of *vestalis* ; but there are slight differences in the relative dimensions of the parts, and the inner margins of the forceps are less hairy, especially towards the base.

I took *Ps. distinctus* plentifully at Colinton, near

Edinburgh, in 1895; it has also been found freely around Rugby by the Rev. F. D. Morice, and in the neighbourhood of Belfast by Mr. H. L. Orr. In East Kent, however, I have taken no specimens except one giant male, nor have I found it on collecting excursions in West Kent, Sussex, and Surrey. Probably, therefore, it is rare in the south-east of England, but common in many places in the midlands and north.

It has been taken in Cornwall, and a male of the yellow-tailed variety has been sent me by Mr. Cunningham from Cargill in Perthshire.

I have seen examples (all white-tailed) from several places in Ireland.

One would expect *Ps. distinctus* to breed in the nests of *Bombus lucorum*, and this supposition is favoured by the fact that at Edinburgh and Belfast *B. lucorum* is much more abundant than *B. terrestris*. To settle the question Mr. Orr kindly sent me a living *distinctus* queen from Belfast on June 6, 1911, and I introduced her to a small *lucorum* colony whose queen was dying. She made herself at home immediately, and subsequently produced 48 young, 34 being males and 14 females; the females were produced after the males.

It is remarkable in how many details the differentiation between *Ps. vestalis* and *Ps. distinctus* corresponds with that which exists between their respective hosts *B. terrestris* and *lucorum*; for

instance, in the tint and tendency to spread of the
yellow, in the length and evenness of the coat, and
in the size of the queens.   In each case the differences are evidently of the same nature and, it can
hardly be doubted, have been induced by the same
causes.

4. PSITHYRUS BARBUTELLUS, Kirby.

QUEEN.—Of medium size; length 17-19 mm., expanse 36-39 mm.; sometimes smaller.

Head black, with some dull yellow hairs on top. **Thorax black, with a dull yellow band in front and a narrow one behind.** Abdomen black, with sometimes some dull yellow on the 1st segment, and with **the 4th and 5th segments,** sometimes also the edge of the 3rd segment, **white.**

The 6th dorsal segment has a dull shine, is punctured, and bears a faint raised line down its centre.

The elevations on the 6th ventral segment form a transverse, crescent-shaped mark, which does not nearly reach to the apex of the segment, and has a narrow division in the middle; between the crescent and the apex the segment is clothed with short velvety golden-brown hairs. The crescent-shaped elevation affords an easy means of distinguishing this species from all others.

MALE.—Length 15-16½ mm., expanse 29-32 mm.

Head black, with some yellow hairs on top. **Thorax black, with a yellow band in front and a narrow yellow band behind.** Abdomen black, with the **1st segment more or less yellow**; the **4th and 5th segments white**; the 6th white at the sides, black in the middle; 7th more or less black. The white often extends on to the 3rd segment.

The coloration is therefore very like that of *Ps. distinctus*, except that there is **no definite tinge of yellow on the sides of the 3rd segment**; in very fresh specimens, however, the white area is sometimes faintly tinged at its base with yellow, but this soon fades away.

Coat uneven—rather shorter than in *distinctus*; the hairs slightly stiffer, slightly coarser, and slightly less dense than in *distinctus*.

On the 6th ventral segment, near the apex, there is a slight mound on either side of the middle.

Hind metatarsi more slender and longer than in *dis-*

tinctus ; they are distinctly narrower than the tibiæ, while in *distinctus* they are scarcely so.

Antennæ rather short, a trifle longer than in *distinctus* ; length of flagellum $5\frac{1}{4}$ mm.   The 3rd joint of the antennæ is shorter than the 5th.

Armature distinct.   Tips of the forceps blunter than in *vestalis* and *distinctus*.

Ps. barbutellus is parasitic on *B. hortorum*. Although widely distributed, it is not very common in most places.   At Ripple it is never so abundant as *Ps. rupestris* and *Ps. vestalis*.   In Ireland it has been recorded from Co. Kerry, Co. Dublin, and Co. Armagh.

5. PSITHYRUS CAMPESTRIS, Panzer.

QUEEN.—Medium sized ; length 17-$18\frac{1}{2}$ mm., expanse 35-38 mm. ; sometimes smaller.

In **light specimens** the head is black, with a few yellow hairs on top ; thorax black, with **a wide yellow band in front and a rather wide yellow band**, often darkened, **behind.** Abdomen black, with the sides of the 3rd segment (except at the base), and the **sides of the 4th and 5th segments yellow.**

This species is **very liable to darken** : the yellow band on the back of the thorax disappears first, then the front band on the thorax, and lastly the yellow on the tail.

Coat rather thinner, and the hairs slightly coarser and stiffer than in any other humble-bee. **All the segments of the abdomen are almost bare in the middle.**

6th segment bare and shining, punctured towards the apex.

The elevations on the 6th ventral segment are rounded and wide and together form a V-shaped mark, with a deep division in the middle.

MALE.—Length 15-16 mm., expanse 28-31 mm.

In **light specimens** the head is black, with some yellow hairs on top ; **thorax black, with a wide yellow band in front, and another wide one behind** ; **abdomen yellow**, with the **2nd segment black**, but yellow at the edge except in the middle, the 6th segment often black in the middle, and the 7th segment black. The yellow on the 1st segment is obscure. Cf. the male of *B. latreillellus.*

In **dark specimens** the whole insect is usually **black, with the exception of the 4th and 5th segments** which are **deep sooty yellow**, interrupted in the middle with black. These segments too are sometimes black.

Hairs slightly coarser and slightly stiffer than in any other humble-bee. **The 6th ventral segment is** slightly channelled down the middle and **bears on either side a tuft of long black hairs.**

Antennæ long; length of flagellum $6\frac{1}{4}$ mm. The 4th joint of the antennæ is very slightly shorter than the 3rd: the 5th joint is almost as long as the 3rd and 4th together.

Armature distinct.

A light male from Chobham, Surrey, in the Saunders' Collection, has the whole 2nd segment yellow with the exception of a few black hairs on its sides.

Ps. campestris is widely distributed in England, Wales, Ireland, and Scotland, and common in a few places. In the Dover district it is seldom seen.

Most authors state that it is parasitic on *B. agrorum*, and its distribution points to this. In Styria, Hoffer found it breeding in the nests of *B. agrorum* and *B. helferanus*.

Ps. campestris, like *Bombus ruderatus*, is subject to colour dimorphism, most examples being either very light or very dark.

Judging by the large number of specimens I have seen in Irish collections this is the commonest species of *Psithyrus* in Ireland. In the lightest Irish specimens the thorax is yellow with the exception of a few black hairs in the centre. None of the numerous black specimens of both sexes from Ireland that I have seen have lacked yellow, or a trace of it, on the tail.

6. PSITHYRUS QUADRICOLOR, Lepeletier.

Synonym :—barbutellus (Kirby), according to Smith.

QUEEN.—Small; length 14-16 mm., expanse 32-35 mm. ; sometimes smaller.

Head black, the hairs on top sometimes tinged with yellow. Thorax black, with a yellow band in front; sometimes a few hairs on the back of the thorax are tinged with yellow. Abdomen black, with more or less yellow on the 1st segment, and with the 3rd and 4th segments white or yellowish-white, and the 5th segment dark reddish-brown. The base and middle of the 3rd segment are generally more or less blackened.

The tip of the abdomen is more incurved than in any other species, so that the abdomen, viewed from above, is shorter and more rounded. 6th dorsal segment slightly and longitudinally raised in the middle towards the base, and slightly hollowed near the apex, where it is coarsely punctured. The 6th ventral segment is narrow and has its apex produced into a sharp hook-shaped point; the ridges on the sides of this segment are small and inconspicuous and do not nearly reach to the apex.

MALE.—Length 13-15 mm., expanse 27-30 mm.

Easily known from any other male humble-bee by the fact that the white (or yellow) tail changes at the tip, generally through more or less black, to reddish.

Head black, with generally some yellow hairs on top. Thorax black, with a yellow band at the front and sometimes a few yellow hairs or a faint yellow band behind. Abdomen black, with generally a faint yellow band on the 1st segment, and with the 3rd segment (except sometimes at the base), the whole of the 4th segment, and often the sides, sometimes the entire area of the 5th segment white, sometimes yellow ; remainder of the 5th segment black, 6th segment reddish or black, 7th segment reddish.

The 3rd and 5th joints of the antennæ are almost
equal; the 4th is about two-thirds as long as the 5th.
Length of flagellum 5 mm.

Armature distinct.

Parasitic on *B. pratorum*, and also, according to
some authors, on *B. jonellus*. It is widely distributed, but does not appear to be common in many
places. At Ripple it is rare. It appears early, the
females in April and the males in June and July;
the latter are fond of the flowers of the raspberry.

In Ireland it has been taken by Mr. Freke at
Borris, Co. Carlow.

On the Continent, in addition to the six species of
Psithyrus found in Britain, there is a species closely
related to *Ps. quadricolor* named *globosus*. This
species is parasitic on *B. soroënsis*, and it is possible
that it may be found in Britain when the haunts of
B. soroënsis are better known. In size and structure
Ps. globosus closely resembles *Ps. quadricolor*, but the
colouring of the coat is like that of *Ps. rupestris*.

IX

ON MAKING A COLLECTION

NOTHING gives greater help and pleasure to the student of any group of insects, whether butterflies, beetles, or bees, than to make a collection of specimens of the different species. To preserve the form and colour of such insects it is only necessary to dry them and to keep them as much as possible in darkness. The British humble-bees make a very pretty little collection. It is not needful to buy an expensive cabinet to hold them. A $14'' \times 10''$ store box, sold by dealers in entomological apparatus for about four shillings, is quite large enough to take about a dozen specimens of each species. If room is wanted for more—and it is a great advantage to have a good series of each sex—another box may be added. Also it is not necessary to set the specimens, that is, to spread out the wings and legs on a setting board, leaving them there until they dry ; though it certainly improves the appearance of the collection, and conveys a better idea of the characters of each species.

Humble-bees are easily caught in an ordinary

butterfly net, though males drowsing on flowers may be picked up in the fingers.

The best way to kill the bees is to place them in a wide-mouthed bottle, in the bottom of which a mixture of cyanide of potassium and plaster of Paris has been spread to a depth of an inch or two. Any chemist will make up this kind of killing-bottle. Care is needed to prevent moisture from getting into the hairs of the specimens lying in the killing-bottle, for this soon spreads over the coat and quite spoils the appearance of the specimen. The moisture is usually caused by the bees regurgitating the nectar they have collected, and is quickly communicated to all the specimens through their rolling about in the bottle. The best preventative is crushed tissue paper put into the bottle before going out collecting; this will absorb the moisture and will keep the specimens from moving about. The bottle should not be exposed to sunshine, for this heats it and in cooling moisture condenses on the inner surface of the glass. Of course the fewer specimens there are in the bottle the less likely they are to damage one another.

A properly prepared cyanide bottle, if it is kept well corked, lasts for a year or two, and renders any bee placed in it motionless and insensible in less than half a minute. If the specimen is to be set, it should at this stage have its wings pulled so that they rest in a horizontal, not vertical, position. It should then be left in the bottle for an hour or two; if less it may revive, and if longer it will stiffen and become hard to set. Some collectors who set their specimens, after rendering them insensible with cyanide, kill them with the fumes from burning sulphur, as this does not stiffen the muscles.

The specimens are pinned through the centre of the thorax. Stout entomological pins about $1\frac{3}{8}$ in. long are best for humble-bees. Only one size of pin should be used for all specimens, great and small, and only about a quarter of the length of the pin should project above the thorax.

The value and interest of the collection are much enhanced by attaching to each specimen a label showing the locality and date of capture. This label should consist of a small square of stout paper run on to the bottom of the pin.

In a freshly-caught male specimen the armature is easily extracted with the point of a pin or needle, and may be either left protruding, or detached and mounted on a card by means of a little liquid glue, the card being afterwards run on to the pin that carries the specimen. Both methods, however, disfigure the specimen unless great care is taken, and seeing that an expert can generally name any specimen without examining the armature, its extraction is not advised except as an aid to the beginner.

In setting specimens the queens should be left on the setting boards for a month, but the males are usually dry and rigid in ten days.

The box or cabinet in which the specimens are to be preserved should be insect proof, and it should contain a small piece of naphthaline.

It requires much time and patience to obtain a
series of perfect, unfaded specimens of some of the
less common species, especially of the queens, by
catching them in the fields, but if a nest can be
found they can easily be bred.   The bees must be
allowed to remain in the nest for three or four days
after they have acquired their full colour or the
colour will fade when they are drying.   The queens
of some species, such as *B. helferanus* and *B.
muscorum*, fly from the nest before their full brightness has been acquired, and in order to breed satisfactory specimens of them it is necessary to keep them
confined to the nest by means of a queen interceptor,
or to bring the nest indoors shortly before it breaks
up, and keep it in a box covered with wire-cloth.
The colours of perfectly mature specimens are
permanent, provided they are not exposed for long
to strong light, the only possible exceptions that I
know of being the delicate shades of lemon and
greenish-yellow in *B. muscorum* and *distinguendus*,
which I have not yet succeeded in preserving in
their full freshness for more than a few months.

There is no better place for collecting humble-
bees than a large old-fashioned garden.   Of the
scores of cultivated flowers that they delight in,
one cannot do more than mention a few : in the
flower garden, nasturtium, sweet-pea, snapdragon,
lavender, bergamot, *Clarkia* and *Fraxinella* ; in the
kitchen garden, sage, broad-beans, scarlet-runners,
globe artichoke (I have seen as many as fourteen
humble-bees on one flower-head of this plant), also

raspberry and fruit blossom of all kinds ; and among flowering shrubs and trees, red ribes, *Escallonia*, *Cotoneaster*, and horse chestnut. In August, when their numbers are greatest and wild flowers begin to fade, they flock to the fields of red clover, and the lazy males congregate on the thistle heads.

X

ANECDOTES AND NOTES

In the following pages are collected some extracts from my notes of various experiences with humble-bees.

"NUMBER 30."

The artificial nests, under Sladen wooden covers, put out to attract queens in 1910, were placed in all kinds of positions, and one of them, No. 30, was situated on our lawn close to where we usually sit out and take afternoon tea.

On June 11, a *lapidarius* queen was noticed going in and out of the nest by my wife, who was sitting only two feet from the entrance. The queen was seen to fly out of the nest at 3.30. She returned at 4.0, and flew out again at 5.5; she returned again at 5.31, and left again at 6.0; she returned at 6.29 laden with honey, but with no pollen, and was seen to go off again at 6.45. Almost every day after this we used to see this queen passing in and out of her domicile when we were having tea, and she showed no alarm at our presence or at the sudden appearance of table and chairs or even at the edges of the table-cloth waving in the wind. Frequently she was seen entering her hole heavily laden with pollen. On June 15 the tea-party was larger than usual, and when the queen arrived on the scene she circled about for a few seconds regarding the company, then she went straight into her hole. At dusk, when she could not see the way in, I sometimes saw her raise herself above the grass and take her bearings from the white wooden cover; with this in sight she flew to the entrance and let herself drop into it.

On June 16, at 11.15 A.M., the queen being out, I lifted the cover and examined the nest. Like most of my other queens this one had ignored a little cavity I had made in the centre of the nest material, and had formed her nest on the top of the material. It consisted of fine material, neatly woven together, with the entrance at the side. The honey-pot, only as yet half its full size, was situated as usual in the entrance, and I could see the honey glistening in it. I separated the little nest from the rest of the nest material, which had become damp, and placed it on some fresh dry material, underneath which I had laid a disc of tin. I also removed all the slugs and centipedes I could find in the cavity, and I cleared the tunnel with a stick, killing two slugs in doing so. Scarcely was this finished when the queen returned, and although I got my garden boy to wave his arms to frighten her off, and both he and I then stooped down over the nest, she took no notice of us, but, seeing her hole, steered straight for it and went in.

It was clear she no longer feared the presence of man.

On June 19, at 10.30 A.M., I placed one of the tin mouse-excluders, previously described (page 119), containing a slot only just large enough for her to pass through, over the mouth of the tunnel. At 3 P.M. she was seen to enter the slot without the least hesitation.

June 21, 5 P.M. Another tea-party. Arriving on the scene she disregarded us, and flew between the legs of a chair into her hole.

On the other hand a passing *lapidarius* queen took notice of an old bag lying in the grass in the paddock near by, and flew backwards and forwards over it as if she was much annoyed with it. Then she passed on. She may have been a queen from one of my other *lapidarius* nests, six of which were situated within thirty yards.

June 26. The queen was seen flying to her hole with her tail inclining to the left side, as if she had met with an accident. I inspected her and found that the tip of her right front wing was gone, and that in consequence she had to turn her tail to the left in order to steer straight when flying with a heavy load.

I again cleared out the nest cavity. It contained a number of young millipedes, and about twenty very small white earwigs.

June 27. The weather having been cold and stormy in the evening, I thought it advisable to see if I could give the queen some food, as I knew she could not have gathered enough during the day to keep her animated through the night. On lifting off the cover at 7.5 P.M. I found her animated, and rather excited at being disturbed, buzzing a good deal, and going in and out of her nest. By pressing back the material over the entrance to the nest I caught sight of the honey-pot which was almost empty. I then, with a fountain-pen filler, injected some syrup which had been prepared for feeding honey-bees, into the honey-pot. As I was doing this I could feel the queen angrily biting the tip of the filler. Directly I withdrew the filler I saw her tongue in the syrup sucking it up, and it remained in this position for quite half a minute. Then the queen retired to her brood, but occasionally I saw her head stretch towards the honey-pot to take a sip. She seemed greatly gratified at finding her pot full, and sipping from it seemed to be a pleasure which she liked to repeat.

June 28. Another stormy day. I filled the honey-pot at 7.50 P.M. The queen was less alarmed than on the previous evening, and drank the syrup eagerly, taking it from the filler.

June 29. At 7.20 P.M. I went to feed the queen. On lifting the cover I was alarmed to see three or four ants (*Lasius niger*) crawling on the nest material. Investigation showed that others were running up and down the tunnel and in and out of its mouth, and even over the honey-pot into the nest. Having had the young brood in two of my nests, one of them but three yards distant from this,

destroyed by *L. niger* only a few days before, I saw
that the ants must be got rid of at once if this nest
was to be saved.   Accordingly I killed eight ants
crawling on the nest material, then caught the queen
in a glass jar, and lifted out the nest.   There was
only one ant in it, but I killed nine more in the
cavity and tunnel and around the mouth of the
latter.   Then with my trowel I shaved off the surface
of the ground—it was grassy—over an area extending to about a foot around the mouth of the tunnel

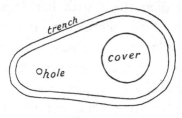

FIG. 32.

and cover, and with my pocket-knife made a little
trench about an inch wide and an inch deep surrounding these, and nowhere approaching nearer to
them than three inches.   Into this trench I poured
a mixture of turpentine and paraffin oil, the scent of
which is distasteful to ants.

   After having made sure no ants remained inside
the area enclosed by the trench, I put back the nest,
placing it, without any superfluous nest material, on
a disc of sacking which covered the disc of tin.
Then I let the queen run in.   She was very pleased
to get back to her brood (the first larvæ had spun
their cocoons), and to find her honey-pot again full.

   June 30, 7 A.M.   No ants in the nest or within

the area surrounded by my little trench. Three dead earth-worms lay in the trench. The turpentine odour had passed off but that of the paraffin remained.

8 P.M. When I came to fill the honey-pot I found that the lump of comb had rolled almost off the sacking, so I hollowed the latter in the middle to retain it. The queen seemed to consider the brood to be insufficiently covered and ran about, pulling and detaching bits of nest material with her jaws and carding them with her legs. She even tried to bite little pieces off the edges of the sacking. While thus occupied she frequently returned to the brood, and always when she reached it emitted little buzzes of pleasure.

July 2. The queen has continued to fly with her tail inclining to the left, but to-day less markedly than at first because both her wings have now become much lacerated. Her flight has become much less swift and strong than formerly. To-day she could fly only slowly and evidently realised her growing feebleness, for she worked a good deal in the garden. She travelled to the sage, *S. officinalis*, knowing well where the two clumps of it in different parts of the garden were situated, and her thorax was dusted with pollen from an orange lily that stood near her nest. She also visited the birds-foot trefoil on the lawn. But she alighted too on grasses and on the flowers of a sow-thistle, from which, of course, she could obtain nothing. How easily she would now have fallen a prey to a bird had it cared to pursue her! She was almost past work, and taking pity on her I provided her with a worker that emerged yesterday in one of my more advanced *lapidarius* nests. About noon I saw her enter her nest, but only a minute later she came out again, and I saw that her honey-pot was already full. Certainly feeding had not made this queen lazy. This, however, was the last day on which I saw her fly.

July 4. In the evening I poured some more paraffin oil into the trench, the smell of the previous application having almost disappeared. No ants had got into the nest.

July 5. The honey-pot having proved leaky, I substituted an artificial one made of beeswax. The first worker emerged to-day.

I continued feeding the colony each evening until the weather improved towards the end of July, but it did not grow very rapidly, and the workers got lazy. In August, however, it picked up well and became quite strong. About the middle of September the weather grew cool and the colony lapsed occasionally into torpor. As time went on the periods of torpor became more frequent and lasted longer. During these periods the bees were able to support life for days together without food. And so the colony lingered on until October 16, the queen remaining alive with four workers until the end.

A FOSTER-MOTHER.

Searching queens caught and put into commencing nests of their own species behaved very variably. A *lapidarius* or *terrestris* queen caught early in the season generally took little or no notice of the brood, even if the nest contained workers and was queenless, but moped in the corner, or if she was permitted to fly she disappeared. But queens caught late in the season sometimes took to the brood at once.

On June 27, 1910, I caught a searching *lapidarius* queen and put her in a box with a cluster of cocoons containing pupæ from one of my *lapidarius* nests that had been deserted ; but she tore them open and would not sit on them.

On July 1, 1910, I captured a searching *lapidarius* queen in the apiary and gave her a few cocoons containing larvæ. She immediately adopted them and showed great attachment to them, sitting on them almost continuously.

On the same day, a *lapidarius* queen occupying one of my domiciles was found in a dazed condition on top of her nest, which had been pulled to pieces and reconstructed by some small mammal, although, and this was strange, her brood which had reached the cocoon stage had not been injured. I brought the poor queen indoors with her brood, but although she drank some honey I offered her, she appeared to be ill, for she paid no attention to her brood but crawled slowly in a stupefied manner around the box with her antennæ stiff and drooping. A recently emerged worker that I had brought from another nest seemed puzzled at her strange behaviour, and followed her about trying to nestle under her.

July 2, 8 A.M. As the sick queen seemed no better and still showed no inclination to tend her brood, I put her and her brood into the box containing the queen caught on July 1 and her brood, and watched the result. The caught queen paid as much attention to the new brood as to her own, sitting now on the one lump and then on the other. She was at first inclined to fear and also attack the sick queen. Discovering, however, that the latter took no notice of her, but only wandered aimlessly about the box, she left her alone. Nevertheless, she uttered a short impatient buzz every time she met the sick queen; it reminded me of the cry often made by queens when they find their brood unexpectedly, but it was a jealous buzz rather than one of pleasure. It was interesting to see how the care of the brood had stimulated her intelligence to its highest pitch of activity. She was keen, alert, and full of life.

At times she endeavoured to leave the box, but one could see by her behaviour that these attempts were not ordinary efforts to escape but were accompanied by an overwhelming consciousness of the presence of her brood, showing that her intention was probably to gather food for it. To prove this, I carried the box containing the nest into the drawing-room and, after closing the windows, opened it. She soon took wing, and after circling round the spot flew to the flowers on a geranium plant in the window, and afterwards to an orange lily in a vase. To the latter she paid particular attention, hovering from floret to floret, dusting herself with the pollen, and, with her tongue extended, rubbing her dangling legs together. A queen humble-bee is too heavy to dart about like a honey-bee, and it was very pretty to watch her manœuvring skilfully and gracefully among the blossoms, every turn executed with intelligence, deliberation, and poise, accompanied by the loud musical hum.[1]

She did not beat herself against the window-panes like a humble-bee usually does that flies in by accident through an open window; this was because she was not frightened and had her interests in the room. Soon she began to search for her nest; I placed it near her and she ran in. Probably it would be easy to train a humble-bee to rear her young in a room or conservatory,[2] while allowing her liberty of flight inside.

10 A.M. The foster-mother was seen feeding the larvæ with liquid food from her mouth. The sick queen had become weaker and crawled slowly about in the lobby apparently oblivious of everything, her only desire being to creep to the light. Her

---

[1] Edison has declared that man is not likely to gain complete mastery of the air until he has learnt to imitate the flight of the humble-bee.

[2] Some years ago the remains of a populous nest of *B. lucorum* was sent me from the palm-house at Kew.

companion occasionally approached her and buzzed, but was afraid to tackle her.

8 P.M.   The foster-mother has been sitting devotedly on the brood and has added a wax rim to the bees-wax cell that I have provided for feeding. The other queen died during the day.

I placed the nest out of doors under one of my wooden covers, giving it six recently-hatched, good-sized young workers from another nest.

July 3.   At 7.30 A.M. the foster-mother was out. I looked into the nest again at 8.0, 10.0 and 10.30 A.M., but did not see her.   No doubt she flew out early in the morning and never returned.   I had half expected this result, for it had happened under similar circumstances in previous years.   The reason seems to be that in nature the queen always learns the position of the nest before she commences to sit on the brood, and when the sitting stage is reached the power to learn a new location is usually lost.   But not always, for I once knew a *terrestris* queen to return.   Although, no doubt, the present *lapidarius* queen carefully marked the position of her home when she left it, the memory of it was soon effaced in the absorbing work of gathering food for the brood, and so she got lost.

11.30.   Happening to be walking in the apiary I saw, searching about in the same place where I had caught the foster-mother on July 1, a *lapidarius* queen, with pollen on her legs.   As an ordinary searching queen is never laden with pollen I watched her for a few minutes, and, concluding from her behaviour that she had no nest there, I caught her and brought her to my motherless nest, a distance of about 300 yards. Thinking she might possibly be the lost queen, I put her, under my net, to the mouth of the hole. She immediately recognised it, and after making several attempts, succeeded in getting through the mouse-excluder I had placed over it and disappeared inside. Two minutes later I lifted the cover and found that she was sitting on the brood, and had unburdened herself of the pollen. There was no question of her identity, and in order to avoid losing her again I caught her in a glass jar, brought her indoors and clipped her wings, afterwards returning her to the nest. Fearing she might stray on foot, I substituted a queen-interceptor for the mouse-excluder.

July 4. The queen spent much of her time this morning trying to get through the interceptor. I visited the nest six times, and every time she was there. At 2.45 p.m., therefore, I removed the interceptor and watched her. After cleaning herself and making two attempts to fly she ran back down the passage quite content. Subsequently the hole was watched for three-quarters of an hour, during which the queen was seen to come out six times, but she never went farther than a foot from the hole, and always with great pleasure found her way back. The interceptor was put back over the hole at night.

July 5. The queen was no more seen trying to get out: letting her have her liberty yesterday seemed to have made her content to remain inside.

From this date the colony prospered exceedingly, and in the middle of August, when its population was at the highest, it must have contained over 200 workers. I attribute this good result to the introduction of a vigorous queen into a nest containing plenty of well-developed worker brood and interfering with it as little as possible. No doubt, too, the earthen domicile under the wooden cover quite suits the requirements of the humble-bees.

AN OBSERVATION NEST IN MY STUDY.

On July 4, 1910, one of my *lapidarius* queens that had started nesting in one of my artificial domiciles was found dead inside a mouse-excluder I had placed over the mouth of the hole. She left a brood of pupæ, larvæ, and eggs, and five puny workers. All these I gave to a searching *lapidarius* queen that had been caught the previous day, and had since been kept in confinement.

I first introduced the queen to the brood. While she was yet an inch away from it she suddenly abandoned her ordinary dull and careless manner and, standing at attention, stretched out her antennæ. No doubt she had smelt the brood and was anxious to get to it, but restrained herself for fear of encountering a rival. Then she cautiously advanced, and when, half a minute later, she reached the brood she showed great satisfaction and pleasure, and immediately stretched herself over it.

After this I dropped the workers, which had become lethargic, into the box : one of them fell close to the queen, who lifted her leg as a warning to it, but when, a few minutes later, the workers recovered animation, mutual friendship was established.

The weather being unfavourable and the workers few and undersized, I did not allow the bees their liberty, but kept them caged and supplied them morning and evening with honey and pollen, and they prospered well.

July 7.   The queen laid a number of eggs during the night.

July 10.   She laid some more eggs.

July 14.   Possessing another caged nest of *lapidarius*, I exchanged queens.   They took to one another's brood and were received amicably by the workers, but at first seemed apprehensive that something was not quite right.

July 17.   I reinstated each queen in her own nest ; this exchange was also effected without incident.

July 19.   The box containing the nest, which was situated on a table in my study, had been accidentally left open, and a worker that had evidently escaped from it was seen to fly in through the open window, laden with pollen.   She flew to the nest-table, which was about three yards from the window, and alighting, ran into the nest.   It was a prettily executed feat, and as the time had now come to let the colony have its liberty, the workers having

increased to twelve, and the weather having somewhat improved, I decided to leave the nest where it
stood on its table in the study, and to let the workers
fly in and out through the open window.

Although this window occupied but one-fourth
of the windowed area of the room, and was opened
only at the top to an extent of about one-third of its
height, the workers seldom flew against the glass.
The window was closed in the evening when all the
bees had returned, about 7-30 P.M., but earlier on
cold days, and later on very warm ones.   The nest,
which was provided with an earth-floored vestibule,
was shut up at the same time.   Next morning
about 8.0, if the weather seemed promising, the
nest and window were opened, but if there was a
high wind the bees were kept confined to their
quarters, and when necessary were fed.   The colony
flourished well, barring an accident to be mentioned
later, and in August, when the weather grew warmer,
it became entirely self-supporting, and reared a
number of males and queens.   No doubt it would
have done still better if the opening in the window
had been nearer the ground—it was 8 feet above
it, for in a gale of wind it was only by exercising
great skill that the workers were able to steer in,
and many must have got battered and lost.   It was
surprising how well they found their way in the
subdued light of the room.   Sitting at my desk I
was almost in their line of flight, but they never
molested me.

Of course it was very easy and pleasant to make

observations on this nest, and I frequently watched the workers dispose of their loads.

On July 20, for instance, I saw a worker cross the room with a full load of pollen. After she had entered the nest I lifted off the cover and saw that the queen took great notice of her, stretching across to her and stroking the pollen with her antennæ. Other workers also showed pleasure and interest in the new arrival, or, as I think more likely, her pollen, and as she went about searching for a cell she produced quite a stir amongst her immediate companions. She buried her head into a pollen cell, then, finding it satisfactory, stepped forward and rubbed the two pellets off her legs into it. Another worker at once dived into the cell and spread the pollen in it.

At 1.45 P.M., the weather having become rainy and windy, a worker was seen busily working backwards and forwards between the nest and a honey-bees' comb of honey that was lying on a lower shelf of the same table, not a bad occupation for a wet afternoon ! At 4.50 P.M. she was still busy at this, and timing her, I found that she took 1 min. 25 secs. to find the comb—this was probably because the room was rather dark—and 4 mins. 50 sec. to fill herself, then she flew straight to the nest. Here she remained $2\frac{1}{2}$ mins. On the next journey $\frac{3}{4}$ min. was spent in looking for the comb, and 4 mins. in filling herself. Three minutes later she appeared again ; this time she flew straight to the comb and took $4\frac{1}{2}$ minutes to fill herself.

July 21. A sad change came over the colony to-day. I noticed some of the workers butting one another this morning. This evening these quarrelsome spirits had multiplied and had thrown the orderly and contented colony into confusion and anarchy. They rushed madly over the comb, attacking and pushing one another with great vehemence, and half their comrades were on the floor idle and drowsy. In the hope of restoring order I removed five of the worst, but still the remaining ones could not agree. Looking at the queen I noticed that she was taking no interest in anything around her but had fallen into a kind of stupor. When I tried to arouse her she trembled, and she would not take food. Evidently she was ailing. A worker seized one of her hind legs in her jaws and began pulling it, another tugged at one of her antennæ, and a third caught hold of her tongue and tried to drag her along by it. She did not seem to mind this rough treatment, but slowly cleaned her antenna with her foreleg when the worker let it go. Probably she was only half conscious. Evidently the object of the workers in seizing her was not to attack her but to remove her; they seemed to have decided that she was useless and going to die.

The sickness of the queen was, in fact, the cause of the quarrelling. In every nest under my observation that has had the misfortune to lose its queen, including one of *latreillellus* and another of *muscorum*, the workers quarrelled in the same way,

R

commencing to do so within an hour or two after the queen was removed or began to sicken. They fight to become mothers. On July 1, 1911, I removed the queen from a strong *lapidarius* colony. Two days later I killed and dissected several of the most pugnacious of the workers; they all contained eggs. This colony settled down somewhat after a day or two, several workers having gained the coveted position, but desultory quarrelling continued, and that it was sometimes carried to the death was shown by the presence of fourteen dead workers (out of a total of about sixty) in the nest on July 15.

To return to the nest in my study. On July 22 at 6.30 A.M. the queen was seen lying dead in the vestibule, no doubt dragged there by the workers. Later in the morning I examined all my outdoor *lapidarius* nests, and found that one that was reigned over by a satellite, the original queen having died, was the least prosperous, this having been the nest that I had assisted least during the prevailing bad weather. I therefore brought this nest indoors and united it to the queenless one. The queen flew out of the cavity, but I gently beat her down with a piece of card into the grass and popped a jar over her; I brought her indoors and let her fly to the window, where I clipped her wings without exciting her, and then put her into the nest. She was pleased to find herself in so large and prosperous a colony, and after running over the brood and hugging it she went to a honey-

pot and took a long draught. All the workers that
approached her seemed interested in her, waving
their antennæ in her direction or touching her, but
none attempted to attack her, and they soon seemed
to regard her presence as a matter of course; they
also ceased to butt one another and peace was re-
stored. The workers showed greater attachment
to the new brood than to their own, no doubt
because it consisted of pupæ in a further advanced
stage. The workers from the outdoor nest, four
tiny ones, were dropped into the nest five minutes
later and were well received.

July 23. The returning workers sweep in
through the window and descend almost to the
floor, then they make a "bee line" for their table,
seeing which they rise and in nine cases out of ten
at once alight and run into the nest.

July 31. The workers, which now number about
twenty, have been lately busy forming a waxen
covering over the nest. During the night $1\frac{1}{2}$
square inches of this were constructed.

August 2. When a worker flew to the window
and found it closed it had the good sense to fly
back again to the nest.

When the nest was accidentally jarred or
breathed upon there was a tremendous buzzing,
that of the queen being the loudest and longest
continued. At the same time a number of the
workers ran out of the nest. When the commotion subsided these workers were in no hurry
to return to the nest, but stood for a long time

silent and motionless in the places to which their excitement had led them, waiting patiently for any further signs of attack. A similarly-caused commotion in a nest of *B. helferanus*, which appears to be a less excitable species than *lapidarius*, subsided much more quickly.

August 6. The workers boldly investigate anything strange, and if I hold a stick to the comb will crawl up it and on to my fingers and hand. They much prefer my hand to the stick or to any articles I may present to them, and it is hard to coax them to leave it. So far I have not been stung, but they behave as if aware they are crawling upon a living being, fearing to venture far, and keeping on the look-out for a sudden attack. I offered pollen to a worker standing on my finger this evening, and she ate it greedily.

Aug. 3. In the evening I saw a worker attack a large open egg-cell containing three eggs. I watched it seize an egg in its mouth and eat it quickly, evidently enjoying it. It then ate another. Another worker ate the third. The queen was engaged elsewhere.

Aug. 4, evening. The egg-cell was larger and sealed, but a worker soon came and made a hole in it. I enlarged the hole and found four eggs in the bottom of the cell. A worker was on the point of lifting out one of these when the queen walked up and sealed up the cell.

Aug. 5. No trace of yesterday's egg-cell was seen this morning at 7.30.

At 3 P.M. the queen had laid the foundation of a large new cell. The workers were already inclined to interfere, shaking their wings excitedly as they approached it; the queen, however, took no notice of their opposition and only went on with the cell at intervals.

I was unable to look into the nest again until 8.30 P.M. The cell was then closed, the eggs having been laid in the interval. It was larger than the cell of yesterday, being about the size of a pea. The workers were very active, trying to open it, but the queen came and pushed them away, butting them severely. I saw her chase a particularly aggressive one, and while she was thus engaged another ran to the cell, and, eagerly opening it, began interfering with the eggs, which, however, being at the bottom, were not very accessible. The queen, returning, hurriedly pushed this worker aside and sealed up the cell. Directly her back was turned two workers began to open the cell again. Losing patience she gave one of them a furious butt which sent it reeling to the bottom of the comb, the other she engaged as she would a rival queen, embracing it as if to sting it, but it easily slipped out of her grip and ran away much frightened and subdued.

A minute later, I looked and saw the queen working with her jaws at the cell in a great state of excitement and activity. Now, instead of closing the cell she had it wide open, and with feverish haste she seemed to be building it up and widening its mouth; at the same time her body revolved round the cell, so that no worker could approach. At times her abdomen was much contracted, and it seemed as if she was on the point of plunging herself, head first, over the cell. I supposed she was about to lay some more eggs, and so it turned out, for her movements became frantic, and ended in her inserting the tip of her abdomen into the cell. She seemed in a great hurry to get the eggs laid; and even while she was laying them the workers attacked the cell, but they did no harm to it. Five minutes later she had the cell sealed over and was guarding it strictly.

August 6, 8 A.M. Peace reigns in the nest, but the cell has completely disappeared.[1] The waxen canopy was much extended during the night and, no doubt, contains the wax from the cell.

4 P.M. Another large cell has been made in the same position as the last. The queen goes and adds to it occasionally, but spends long intervals crawling over and incubating different sets of cocoons, quite unconcerned. Three or four workers prepare for their campaign of opposition by planting themselves near the cell, where they maintain a continual buzzing and shaking of the wings.

4.30 P.M. The queen is laying. While the eggs are being laid the tip of the abdomen remains in the cell, and the sting appears through the wall of the cell as each egg passes. When the queen lifted her tail out of the cell

---

[1] I do not think that cells of eggs are often completely destroyed in a normal nest. In this nest the workers' passion had been excited by the change of queens on July 22.

three eggs were seen in the bottom of the latter. She was now kept busy repelling audacious workers, but often, as on the previous day, when she attempted to follow one, another would seize the opportunity to put its head into the cell, and make a show of devouring the eggs, although the great depth of the cell prevented its doing any injury to them.

The queen evidently had more eggs to lay, for she did not close up the cell, and the next minute she was laying again. Again she left the cell; again she laid; she was very active. At 4.35 when she took another rest there were eight eggs in the cell. During this interval, while the queen was working at the wall with her jaws, a worker plunged its head into the cell right under her nose. For a moment the queen continued to work as if she had not noticed her, then growing exasperated she seized the worker and pitched it headlong. She laid one or two more eggs, and then, at 4.42, hurriedly sealed the cell over, and commenced guarding it.

At 8 P.M. the queen was still keeping a close guard over her eggs. A worker occasionally succeeded in making a small hole in the top of the cell, but the queen soon discovered the mischief and repaired it.

A worker attacking the cell did no harm to the eggs during the first half minute, but merely, as it slowly bit the cell open, made a great fuss which usually attracted the queen. It was only if the queen failed then to come to the rescue that the eggs were jeopardised.

August 7, 4.40 A.M. Again the cell has completely disappeared. Peace reigns.

3.5 P.M. The foundation of a new cell is begun.

3.20 P.M. The cell wall is already $\frac{1}{8}$ inch high. The workers, which show the usual excitement, are more busy working at it than the queen. They seem even to add wax to it, certainly they take none away. Only one bee works at the cell at a time. At this stage the queen makes no attempt to stop the workers, except on rare occasions when she wishes to work at the cell herself, and then a slight push is sufficient to displace them.

3.40 P.M. About six workers are busy at the cell, taking turns at the work. It grows in height, and the queen still visits it so seldom that it seems clear the workers are building it and not merely interfering with it.

The queen had her first egg laid at 4.10,

,,	second	,,	4.11$\frac{1}{2}$,
,,	third	,,	4.12$\frac{3}{4}$,
,,	fourth	,,	4.13$\frac{1}{2}$,
,,	fifth	,,	4.15,
,,	sixth	,,	4.17$\frac{1}{2}$,
,,	seventh	,,	4.19,
,,	eighth	,,	4.20,
,,	ninth	,,	4.21$\frac{1}{2}$,

and had the cell sealed at 4.23.

She lifted her tail out of the cell between the laying of each egg. On each occasion the time her tail was in the cell was only twenty to thirty seconds. The rest of the time was spent in working at the rim of the cells with her jaws, and often, just previous to each laying, in making one or more futile attempts to insert the tip of her abdomen in the cell, her tail slipping down the wall of the cell on the outside.   An attempt having at last proved successful she grasped the cell with her hind legs, and in a few seconds the tip of the sting appeared through the wall of the cell and remained exposed for a moment or two.   On one or two occasions a worker noticed the sting and touched it with its jaws. The queen began reducing the mouth of the cell after laying the seventh and eighth eggs, as if she was going to close it, but enlarged it again just before laying another egg.   After sealing, the workers as usual became very troublesome.   The queen seized energetically the most persistent of these, and inflicted on them a punishment which seemed to consist of a snap or two with her jaws.

All was well with the cell when I looked at it on three occasions between 5.0 and 7.0 P.M., but at 7.25 I found it wide open, with frenzied workers eagerly biting at the eggs.   The light being poor, I went to fetch a candle, and when I returned the cell was sealed and the queen keeping guard.   A few minutes later I was surprised to see her open the cell and lay three or four more eggs.

Of course the workers continued to show great hostility to the cell of eggs.   In order to ascertain how far the disaffection extended amongst them, every time I saw a worker attack the cell I took it away. In this way I removed ten workers. Still some assaulted it, but their attacks were less vehement.

Aug. 8, 7.30 A.M. The cell remains, but is much reduced in size, and is not nearly large enough to contain all the eggs; some must have been removed during the night. The workers do not now molest the cell, but cherish it, and the queen takes no notice of it.

About 5.0 P.M. a fresh cell was constructed in another part of the comb, and the queen laid a number of eggs in it and kept guard over them. At 7.45 she added one or two more eggs to these, bringing the total number in the cell to about ten.

Aug. 9. This cell has also survived the attacks of the workers.

At 1.20 P.M. another cell was begun. Three or four workers were busy at it, but the queen visited it only occasionally, and at 1.50 it had disappeared; probably the workers intended this cell for the reception of their own eggs.

At 2.45 again a cell was started; this was slowly completed, and the queen began to lay at 3.50. Watching her at work it seemed to me that she often missed the cell with her tail because she was not quite ready to lay, for a determined effort always succeeded.

Aug. 10, 7 A.M. This third cell of eggs has also survived the night, and is not so much reduced in size as the last. Evidently the queen has learnt better how to protect the eggs, or the workers have become less aggressive.

At 5.10 P.M. The queen laid another batch of eggs and it survived.

On Aug. 11, at 6.30 P.M., she laid another; this also survived.

On the afternoon of Aug. 12 another. On the 13th, the usual daily batch was laid at 5.10 P.M.; on the 14th, about 5.0. P.M. Eggs were also laid by her on the afternoons of the 15th, 16th, 17th, 18th, 19th, 20th, and 22nd. Each of these batches was carefully defended by the queen for several hours after it was laid, but on many occasions the workers must have succeeded in destroying some of the eggs; for instance, I saw them at 8.30 P.M. on August 16, busily devouring those laid that afternoon. After the 22nd the queen's prolificness fell off rapidly, and on several days she laid no eggs at all. On the 26th workers were seen butting one another for the second time. This was a sure sign of the failing powers of the queen, but on Aug. 31 at 5 P.M. I found her defending a cell of recently-laid eggs. On Sept. 3 she was found in a drowsy state on the floor of my study, and the next day she died.

OBSERVATIONS ON PSITHYRUS.

On July 19, 1911, I dug up a nest of *B. lapidarius* victimised by *Psithyrus rupestris*, and transferred it to my humble-bee house for observation. The nest contained 71 workers and a large amount of brood. Workers were still emerging.

The comb consisted of:—

(1) The empty cocoons that had contained the first batch of workers.

(2) Three large clusters of later worker cocoons, all empty except four cocoons in one cluster which, by their darkened appearance, showed that their occupants were about to emerge.

(3) Four large clusters of *rupestris* cocoons, none yet hatched. The oldest cluster contained 21 males and 5 females ; the second, 21 males and 6 females ; the third, 18 males and 5 females ; the youngest, 10 males and 9 females.

(4) Three lumps of half-fed and nearly full-fed larvæ. Three lumps of smaller larvæ. Three cells of eggs ; one of these was opened, and was found to contain 29 *rupestris* eggs.

The *Psithyrus* queen was nowhere to be seen, but suspecting she had hidden herself in a side hole, I left a lump of cocoons in the cavity to attract her, and returning a quarter of an hour later found her on this.

To strengthen the nest I added brood and workers from a queenless *lapidarius* nest.

During the next few days I watched the nest closely. The *rupestris* queen was most energetic, and day and night kept running hurriedly over the comb in all directions, uttering a short mournful cry every minute or two ; the sound was like that emitted by a bee when it is squeezed, and lasted about a second ; while it was being made the wings were seen to vibrate. Her chief occupation seemed

to be to chase certain workers, probably those that
desired to lay eggs. When she caught a worker
she clasped it to the underside of her abdomen and
rolled over with it, apparently endeavouring to
squeeze or sting it. But the worker was too small
to get hurt, and soon slipped away. The incurvature of the abdomen, and the ridges under its tip,
peculiar to all *Psithyrus* queens, must be of great
assistance in thus dealing with refractory workers,
and this, I think, is probably their function.

On July 24, at 6.0 P.M., she had laid a batch of
eggs in a cell which must have been constructed
very rapidly, because there was no sign of it at 5.0.
She guarded the cell from the attacks of the workers
in much the same manner as a *lapidarius* queen,
but I noticed she spent a much longer time in giving
the finishing touches to the cell, and while she was
doing this the workers dared not approach.

At 9.30 P.M. she was still guarding the cell from
the workers, several of which had now grown very
bold, and occasionally succeeded in biting holes in
it, the *Psithyrus* having become less vigilant. The
workers shook and revolved their wings in the
greatest excitement while they were attacking the
cell. She carefully sealed up every rent they made.

Next morning (July 25) the cell still stood, and
the workers had ceased to assault it.

At 8.0 P.M. the foundation of another cell was
laid close to the old one. The workers were very
excited over it, nibbling it a great deal. Now and
then it almost disappeared, nevertheless the workers

continued to work feverishly at the site, apparently making a show of attempting to devour imaginary eggs in it. The queen took very little notice of them but pushed them aside when she came to add wax, which she gathered hurriedly, chiefly from the joints between recently-spun cocoons. During the first hour the construction of the cell made very little progress. At 11.30 P.M., however, it had distinct walls.

July 26, 6.30 A.M. There was a loud humming in the nest when I entered the bee-house, and on looking at it I saw at once that something unusual had happened. The whole colony was in a state of uproar, and the workers were rushing about shaking their wings. All trace of the cell that was building last night had vanished. The *Psithyrus* was not to be seen anywhere on the comb, but I soon discovered her in the tube leading to the ground. She was drowsy, and her coat was dishevelled and matted with moisture. I got her out and dropped her into the nest. Immediately the commotion increased threefold, and I perceived that she was the cause of it. The workers were infuriated with her, and half-a-dozen rushed after her and hunted her into a corner where they all set upon her and tried to sting her. She struggled feebly and ineffectually. I rescued her and put her into the vestibule. Here, however, five minutes later, I found the workers had followed her; one of them had a wing of the *Psithyrus* tightly clasped in its jaws and was trying to sting her with all its might.

Seeing a restoration to the workers in their present temper was impossible, I took her away. She was cold and lethargic and refused food. But at 11.15 A.M. she seemed to have recovered somewhat and, after taking some honey which I offered her, she became, to all appearance, quite well and lively.

At 11.30 I decided to put her back again into the nest. The commotion among the workers had subsided to a great extent, and they were beginning to quarrel between themselves, occasionally butting one another; but as soon as the queen was introduced the uproar began again and they rushed towards her and attacked her fiercely. One seized her wing, another took hold of her leg and tried to drag her by it, while several others climbed on to her back and tried to sting her. She made no attempt at resistance, but endeavoured to creep away. Seizing a favourable opportunity, I pushed the glass lid to one side and let her crawl out. Although she had been only about two minutes in the nest, she had in that time fallen into the same state of depression, lethargy, and coldness as when I rescued her before. Seeing that it was of no use to try and reinstate her, I killed her. During the afternoon the workers constructed an egg cell in a new place and laid a number of eggs in it. Throughout the afternoon and evening this cell was the centre of a great deal of quarrelling; first one worker then another held it, single-handed, against all comers. One began to bite it open, probably because it had an egg or two to lay in it, but was thrown down by a terrific charge from another which proceeded at once to seal up the cell. But the latter soon left this work to rush at a third which drew near. Hurrying back it found that a fourth had dared to approach the cell. This one seemed to be particularly dangerous and hateful, and seizing it, it hurled it down and tried to sting it. This was the opportunity for a fifth to take possession of the cell, and this one in its turn jealously and energetically guarded it until it too was deposed. Thus the squabbling went on, but there was no general commotion as when the *Psithyrus* was present, and most of the bees not engaged in laying eggs were peacefully occupied incubating or feeding the *Psithyrus* brood.

It was clear that the workers deposed the *Psithyrus* queen, and I think that this was the culminating act in a revolt that the queen had all along found it difficult to repress. The episode was much more than a mere conflict between a *rupestris* queen and a number of *lapidarius* workers. It was an incident in the everlasting struggle between a parasite and its host, in which the destinies of both species are involved. The *rupestris* queen, following her instinct for the propagation of her species, laid her eggs and endeavoured to prevent the *lapidarius* workers from laying any of theirs. The *lapidarius* workers, on the other hand, endeavoured to destroy the *Psithyrus* eggs, and were ready to lay some of their own. The *Psithyrus* would have triumphed, as she always does in nature, had not something happened to weaken her power and increase that of the workers. That something, I believe, was my adding workers from the queen-less nest. Many of these workers contained eggs, and it was over these that the *Psithyrus* had been unable to maintain her authority.

I think that the strange race-suicidal habit the *lapidarius* workers have of attempting to devour their mother's new-laid eggs is associated with the parasitism of *Psithyrus*. It is natural to suppose that workers that attempt to devour the eggs of their *Psithyrus* step-mother perpetuate their egg-devouring instinct through their sons that they sometimes succeed in rearing. In support of this view it is interesting to note that in nests of *B. latreillellus*, a species that is not preyed upon by any species of *Psithyrus*, I have never seen the queen's eggs molested by the workers.

On July 20, 1911, I took a nest of *B. terrestris* victimised by *Ps. vestalis* and transferred it to my humble-bee house. There were 49 workers, mostly small, 3 *vestalis* males that had emerged within 24 hours, the old *vestalis* queen, a long-dead *terrestris* queen and the remains of another, 67 *vestalis* female cocoons, 103 *vestalis* male cocoons, and a quantity of larvæ and eggs. One of the egg-cells contained 18 *vestalis* eggs. The brood had the same smell as that of the *terrestris* brood in my other nests except that it was rather stronger; the *terrestris* workers seemed exceedingly fond of it, and when I brought

S

it home a worker from one of my *terrestris* nests settled on it and began incubating it.

On July 22, about 9 P.M., I saw the *vestalis* queen occasionally paying attention to two empty egg-cells. Possibly these had contained workers' eggs, for at another time I saw her quietly eating the eggs in another cell. Sometimes she worked at one cell, sometimes at the other. Soon she became busier and began to work pretty continuously at one of the cells, adding wax to it. At 9.54 she commenced laying eggs in this cell without having previously worked herself up into any excitement, putting the tip of her abdomen into the cell and clasping the latter with her hind feet just like a *Bombus* queen. She kept the tip of her abdomen in the cell for about two minutes, and her sting appeared through the wall of the cell several times. During the first minute not a worker approached, but during the second minute she was continually harassed by a worker trying to bite down the wall of the cell from behind. She now kept trying, but without much effect, to beat the worker's head down with her hind feet, which, owing to the incurvature of the abdomen, extend farther beyond it than in *Bombus*. The worker, however, did not excite or hurry her, and, in fact, did no material damage to the cell. When she had finished I saw several eggs in the cell; they must therefore have been laid very quickly. She immediately sealed up the cell and then began polishing the surface of it with her jaws. She maintained this polishing work, almost without ceasing, for at least an hour, which was as long as I watched her. No workers assaulted the cell, but the queen was so busy at it, walking round and round it as she rapidly polished it, that they could not have succeeded had they tried.

This cell was standing next morning.

As time went on more eggs were laid, and a large number of *vestalis* males and females developed; but although the nest contained many fertile workers, not one *terrestris* male was produced.

A CRIPPLED TERRESTRIS QUEEN.

On May 4, 1911, I caught two *terrestris* queens, *A* and *B*, and kept them together in darkness. Their quarters, like those of all my other captive queens, consisted of a nest starting-box communicating by means of a small hole with a vestibule or outer box. On the floor of the nest-box I placed a disc of sacking, to which were firmly fastened by melted bees-wax two deep cells of bees-wax, which I filled twice a-day with a mixture of honey and water; I also fastened to the sacking a shallow cup made of old *lapidarius* wax containing a lump of pollen. *A* had her left hind tibia slightly paralysed, the result of an injury sustained before she was caught.

May 8, 9.0 P.M. *A* formed a cell in the place that the pollen lump, which was now consumed, had occupied, and she laid an egg in it. *B's* tongue was seen to be injured, she could not fold it up properly.

A looked after her brood energetically for several days and laid more eggs, but *B* was generally drowsy. On May 13, however, both queens were incubating the brood. The paralysis of *A's* hind left tibia had increased, and, in consequence of her inability to use this limb, the left side of her abdomen had become clogged with wax.

May 14. At a distance of a quarter of an inch from the brood a waxen cell has been constructed $\frac{3}{8}$ in. in diameter and $\frac{1}{8}$ in. deep. This, I think, was an attempt at a honey-pot, but no honey was ever placed in it.

About this date two or three batches of eggs were laid in cells adjoining the first egg cell. The queens continued to live on friendly terms with one another, sometimes the one sometimes the other incubating the brood, until May 26, when *A* killed *B*. Such a long-continued friendship of two incubating queens was unique in my experience, and I attributed it to the fact that both were maimed.

On May 29 *A's* paralysed tibia, which had been dead and useless for some days, dropped off. An enormous quantity of wax had now accumulated on the left side of her abdomen, and as this greatly impeded her movements, I determined after I had examined it to remove it. The wax had exuded from the base of each of the dorsal segments except the 1st and the 2nd, and formed lumps which extended nearly half-way across the abdomen. The lump at the base of the 5th segment was by far the

largest, and, after removal, made a ball of wax
$\frac{1}{8}$ in. in diameter. Only about one-third of this
quantity had exuded from the base of the 4th
segment ; the amount at the base of the 3rd segment
was equal to the amount at the base of the 4th,
while the quantity at the base of the 6th segment
was very much less, and I computed it to be about
one-tenth of what had oozed from the base of the
3rd segment. The 2nd segment on the left and the
under side of the abdomen on the left were greasy,
but they bore no wax.

I mention the quantities of wax as accurately as
I was able to estimate them, because I think that
they probably represent the proportion of wax that
exudes from between each segment in a normal
queen. It will have been noticed that the quantity
given off between the 4th and 5th segments
was greater than the total quantity from all the
other segments. This is the part of the abdomen that
is most easily reached by the brushes on the hind
metatarsi, which no doubt are used to remove the
wax, for in a queen that had one of her hind metatarsi slightly paralysed the brush on it was covered
with wax.

The fact that my *A* queen, having lost her
left hind tibia, had been able to keep the right side
of her abdomen clean shows that the wax is not
removed from the metatarsal brush by the tibial
comb and passed on to the corbicula like the pollen.
How, then, was it removed from the metatarsal brush
in this case ? At the apical end of the metatarsal

brushes on the middle and hind legs is a row of extra long bristles (shown clearly in Fig. 5, A), which probably act as a comb for cleaning, amongst other things, the brushes on the other metatarsi. In the case of the present queen the wax on the metatarsal brush on the right hind leg was probably removed by the metatarsal comb on the left middle leg.

The removal of the wax greatly relieved the queen, and it did not accumulate again in any quantity : this confirmed what I had long suspected, namely, that the queen humble-bee almost ceases to produce wax after her first workers appear.

Two *terrestris* workers were given to the queen the day before she killed her comrade, and these nursed the brood well. On June 5 the first young worker emerged, a large and perfect one. A second worker appeared on June 8, and others followed on June 21. Needing more room in my humble-bee house, I united the nest to another, the queen of which unfortunately killed my crippled pet.

MISCELLANEOUS NOTES

Towards the end of June 1899 I exhibited two small colonies of *B. terrestris* at the Royal Agricultural Society's show at Maidstone, in a little hive set on a table far inside the bee-tent. The workers easily found their way to their hive although it was surrounded by a crowd of interested onlookers, who were able afterwards to see them depositing

their loads of pollen and honey under the glass.
Their good behaviour earned for them a first prize.
The vibration of the railway journey from Dover
to Maidstone upset the colonies a little at first;
but the queens soon returned to their brood, and,
on the return journey, one of them was so reconciled
to the motion that she laid a batch of eggs in the
train!

When queens grow elderly and their prolificness
declines, they lose their desire to fight one another.
On July 25, 1910, I added a *pratorum* queen from
a declining nest to a strong nest of the same species.
The two queens lived happily together until one of
them died on August 15. Two *lapidarius* queens
from strong nests, put together on August 7, also
remained friendly.

From July 15 to 27, when the above-mentioned
pratorum nest was at the height of its strength, the
weather was very unfavourable, with strong cold
winds, the temperature often failing to exceed 60° F.,
and there was very little sunshine; yet even on the
worst days a fair quantity of honey was gathered
and stored in the cells, although during this period
my honey-bees not only failed to gather honey, but
consumed all previously gathered stores, and had
to be fed to avoid starvation.

The comb and nest of each species have a
distinct smell, although to our imperfect olfactory
sense the differences do not in all cases seem great,
and we have no words to describe them; the bees,
too, have a distinct smell, which is not unlike that

of the comb; this scent may be perceived by smelling the bees that have been collected in the glass jar when a nest is taken, after they have been allowed to quiet down. But if these bees are irritated, by being breathed upon, for instance, another smell, namely that of sting-poison, is produced. That the humble-bees themselves are highly sensitive to these different smells I have had many striking proofs. Once I saw a *lucorum* queen enter one of my domiciles in which a *lapidarius* queen had collected a lump of pollen and deposited her first eggs. Although the *lapidarius* was not at home, the *lucorum* queen bundled out of the nest in great confusion, whether disgusted or terrified I cannot say.

This antipathy to the smell of a strange species and its brood is probably the chief obstacle in the way of getting different species to fraternise in a nest. The introduction of cocoons containing workers near emergence from a *sylvarum* nest into a strong *lapidarius* nest produced a great commotion which did not completely subside for a day or two, and when the *sylvarum* workers crept out they were all killed. A small pellet of *ruderatus* wax placed in the same colony threw it into an uproar.

On the other hand several remarkable associations between strange species occurred. A dwarfed *lapidarius* worker in some way got out of one of my *lapidarius* nests and entered a nest of *terrestris* in which a queen and two workers were confined. She immediately became one of the family, paying as much attention to the brood as the *terrestris* workers did. When the nests in my bee-house were about to break up at the end of the season, the workers of the different species were often found in one another's nests. In all these cases the nests were weak; I have never known a bee of a strange species to gain admittance into a strong nest. Even *Psithyri*, as has been mentioned, are not suffered to enter colonies of the species on which they prey if they are populous.

The brood of the humble-bee is less easily killed by cold than that of the honey-bee. I have known *lapidarius* workers to hatch from cocoons that have not been incubated for two days, and I was sometimes successful in getting bees to emerge from deserted cocoons by placing them between the quilts that covered my colonies of honey-bees.

If I wished to do anything to or with a queen without exciting her, for instance to clip her wings, I found it best to let her fall into a drowsy state by confining her in a box for an hour or two without food.

There are two ways of holding a queen in the fingers so that she cannot sting: (1) by grasping both wings close to the roots;[1] and (2) by gripping the thorax on both sides.

The eyes of a humble-bee, like a photographic plate, are less sensitive to yellow light than ours.

[1] This method is also safe for a worker humble-bee and for a worker honey-bee of the Italian race or of my British Golden breed, but not for a worker of the English black honey-bee, nor for a wasp, the abdomen being more mobile in these.

This was clearly demonstrated by the following incident.   One evening I was examining a nest in my humble-bee house with the aid of a candle, it being too dark for me to see even to read, when the queen escaped.   Instead of flying to the candle she made for the open door.   Yet it is plain that

FIG. 33.—How to hold a humble-bee.
Grasped in this way by the wings she cannot sting.

the flame of a candle is faintly perceived because if there is no other source of light a frightened bee will fly to it like a moth, although it is only by exciting her greatly that she can be induced to take wing.

Several instances, not only of the good eyesight of humble-bees but of the intelligence with which they use their eyes have been given.   I will here add another which seems worth recording.   On May 18, 1911, a *lapidarius* queen from one of my artificial domiciles was seen visiting a bunch of bluebells that had been placed on the sill outside the kitchen window.   A similar bunch of bluebells was standing on a table inside the closed window of my study.   The queen several times came round to my study window and persistently beat against the glass in her endeavour to reach the flowers inside.   This incident shows how humble-bees note the kind of situation in which their favourite plants grow, and how quickly they learn to search for them there.

No one who has watched humble-bees gathering food can doubt that they know and distinguish the flowers by their colours rather than by the scent of the nectar.

They know where all the clumps of their favourite flowers are situated, and may often be seen paying a round of visits to groups of white-dead-nettle and hedge-woundwort.   Not only this, but they frequently remember the position of each plant and even of each floret, for I have often observed them travelling from one half-hidden floret to another with the swiftness and assurance of familiarity.

Under ordinary circumstances, whether she is in or out of the nest, a humble-bee that is in an animated state, that is, whose abdomen is expanding and contracting, always shows signs of consciousness, her head, even when she is unoccupied, being held erect and her antennæ pointing at attention. But occasionally I have found a fully animated queen resting on a plant or flower with her head hanging down and her antennæ resting on her face, evidently indulging in a nap like the sleep of the higher animals, for she awakes with a start when disturbed. Before the cares of motherhood have come upon her, the queen is very fond of dropping off to sleep in the warm sunshine or in her newly-found nest.

We have seen how the legs are employed in collecting pollen and wax; they have also a third and less specialised use, namely, to keep the insect clean. Several pages would be needed to describe the various motions they make in brushing and combing the different parts of the body. It is sufficient to say that the head together with the tongue and antennæ is cleaned by the front legs, the thorax by the middle legs, and the abdomen and wings by the hind legs. In the middle and hind legs it is the brushes on the inner sides of the metatarsi that do most of the work. Every speck of dust and pollen is removed by repeated brush-ings. Finally the legs are cleaned by being rubbed together. The cleaning of the head and antennæ is often the prelude to taking a flight, whereas the brushing of the coat on the thorax and abdomen is frequently performed when the bee returns to the nest after a dusty forage in the fields. Many queens when they grow old get careless about cleaning themselves, their unkempt coats becoming matted and greasy.

In a sudden shower a *lapidarius* or *terrestris* worker will sometimes take shelter under the eaves of a building or the horizontal bough of a tree. I have never seen a honey-bee doing this.

It may be observed that the centre of attraction and affection in a colony of humble-bees is not the queen nor the food, but the brood, the future representatives of the race, and especially that which is soon to emerge.

PLATE VI.

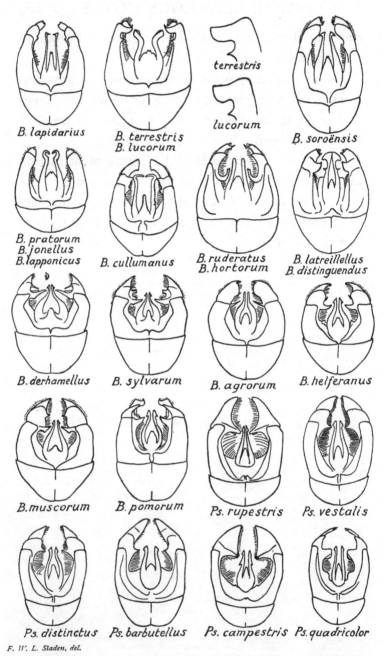

terrestris

lucorum

B. lapidarius

B. terrestris
B. lucorum

B. soroënsis

B. pratorum
B. jonellus
B. lapponicus

B. cullumanus

B. ruderatus
B. hortorum

B. latreillellus
B. distinguendus

B. derhamellus

B. sylvarum

B. agrorum

B. helferanus

B. muscorum

B. pomorum

Ps. rupestris

Ps. vestalis

Ps. distinctus

Ps. barbutellus

Ps. campestris

Ps. quadricolor

F. W. L. Sladen, del.

♂ **Armatures of all the British Species of** *BOMBUS* **and** *PSITHYRUS.*

ADDITIONAL NOTES
MADE IN JUNE AND JULY 1912

The Season of 1912.—Queens of the common species were unusually abundant in the spring of 1912, probably as the result of the favourable summer of 1911. On still, warm days their murmur, searching for nests in woods and copses, was continuous, and it was often possible to guess correctly which species was searching in one's vicinity by its note alone, that of *B. terrestris* being the loudest and most often heard. But although the weather continued very favourable, nests in June and July were scarce, and it was found that large numbers had been destroyed in early stages by *Brachycoma devia*.

Microsporidiosis in Humble-bees.—In the *Report on the Isle of Wight Bee Disease* (*Microsporidiosis*) issued by the Board of Agriculture in May 1912, Dr. H. B. Fantham and Dr. Graham-Smith state that they have found in the chyle stomach and Malpighian tubes of a number of dead and dying specimens of *B. lapidarius, terrestris,* and *hortorum,* picked up under trees at Cambridge and elsewhere in August and September 1911, a protozoal parasite belonging to the genus *Nosema,* resembling *N. apis,* which has recently caused enormous mortality amongst honey-bees in certain parts of Britain.

How the First Eggs are laid (p. 19).—The lump containing the first eggs was carefully examined in several nests in 1912. A nest of *B. lapidarius* had 12 eggs, all in one cell, and laid in a bundle. A nest of *ruderatus* had 14 eggs, each occupying a separate bed in the pollen. Another nest of *ruderatus* had six young larvæ, varying from the size of a lettuce seed to that of a mignonette seed in one cell, 3 eggs in another cell, and 5 eggs in another.

How Pollen is collected (pp. 20 to 24).—Many variegated loads of pollen collected by *Bombi*, consisting of two or more kinds, were examined during 1912, and they showed clearly not only that the pollen is pushed into the corbicula in the way explained on p. 22, but how the load is built up.

About 40 per cent of the loads brought home by a colony of *terrestris* between July 15 and July 25 could be

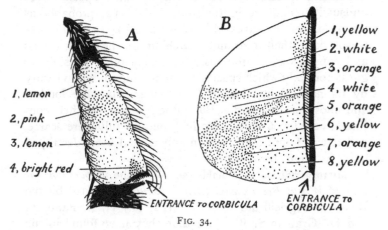

FIG. 34.

A, Shallow load of pollen in the left corbicula of a worker of *Bombus agrorum* captured at dead-nettle on July 4, 1912. *B*, Section through a full load of pollen in the left corbicula of a worker of *B. terrestris* taken on entering its nest, July 23, 1912.

The layers of the different kinds of pollen are numbered in the order in which they were collected. Had the *agrorum* worker gone on collecting and gathered a full load, the pollen shown in its corbicula would have occupied the region referred to as "1, yellow" in Fig. *B*.

seen at a glance to consist of two kinds of pollen; in a few of these, on closer examination, three or more kinds were detected.

Pollen-primers (p. 43).—*B. latreillellus* was considered to be a pollen-primer because pollen was found under the eggs in a nest in an advanced stage examined in 1911.

But in a nest in an earlier stage kept under observation in 1912 the eggs were laid in cells that contained no pollen, although in at least one case pollen was put into the cell and removed before the eggs were laid. Should future investigation show that *latreillellus* is a pollen-primer only under abnormal conditions, a better name for the group, consisting of *ruderatus, hortorum, latreillellus,* and *distinguendus,* would be " Long-faced Humble-bees " ; the term " pollen-primers " could then be restricted to *ruderatus* and *hortorum. Latreillellus* and *distinguendus* are not closely related to *ruderatus* and *hortorum.*

Development of the Queen (p. 50).—I have succeeded in getting queens reared by *B. latreillellus* from eggs laid early in the queen's life in the following manner :—Two *latreillellus* nests, in which a few workers had already emerged from the first batch of cocoons, were joined together, and the queens and all the young larvæ were removed, and also all the eggs except three. The larvæ that hatched from these three eggs had thus about twenty workers to care for them exclusively, and they developed into females as large as queens. That these were really queens and not giant workers was shown by the fact that they paid no attention to the brood that was subsequently reared, and as soon as they were old enough to fly they left the nest for good.

An attempt to breed from Queens of *Bombus soroënsis.*— On May 25, 1912, I received from Mr. E. B. Nevinson at Abersoch two queens of *B. soroënsis,* from which I attempted to breed in the manner explained on pp. 131 and 132. Within three days they had settled down together very contentedly, and on the fifth day it was plain they had eggs, which they incubated assiduously. I gave them two just-hatched *terrestris* workers on June 1, and a just-hatched *lapidarius* worker on June 4. The brood lump swelled rapidly on June 7 and 8, but on the morning of June 9 I was disappointed to find it much smaller and three half-fed larvæ lying in the vestibule. Examination of the lump showed that it contained one egg. (Confined *terrestris* queens sometimes eat their own or each other's eggs.) I now separated the two queens—they had remained good friends—giving to the one a first batch of *derhamellus* cocoons, soon to hatch, and to the other a first batch of *latreillellus* cocoons. Each queen incubated her brood with the greatest attention. In each case the workers that hatched were at first friendly to the queen, but later they turned the queen out of the nest and laid eggs of their own.

Behaviour of Queen of *Psithyrus rupestris* **in Nests of** *Bombus lapidarius.*—On June 13 I put a *rupestris* queen into a *lapidarius* nest containing 14 workers and much brood. The *lapidarius* queen rushed out of the nest and ran about the vestibule in great agitation for a long time. Occasionally she peeped into the entrance, but she hastily withdrew. Once I gently pushed her in, but with extraordinary strength she forced herself back. I next dropped her into the nest, but she dashed wildly about and escaped again to the vestibule. The *Psithyrus*, emboldened by her fear, tried to sting her as she passed her. I afterwards removed the *Psithyrus*, but the *lapidarius* queen did not dare re-enter the nest, and she fell into a drowsy state in the vestibule.

A *lapidarius* nest containing a *rupestris* queen was kept under observation for three weeks, from July 2 to July 23, 1912. On July 2, when the *Psithyrus* was introduced, the nest contained about 30 workers and a great deal of brood, and she killed the *lapidarius* queen on the same day. The *Psithyrus* spent most of her time chasing certain workers over the comb, and thus prevented them forming cells in which to lay their eggs. Between July 10 and July 23 she laid a large batch of eggs almost every day. She built the egg-cells herself with wax she gathered in the nest. The workers became very excited over the egg-cell during its construction. She made her cell and laid her eggs in it very rapidly. On one occasion the eggs, numbering 23, were laid in six minutes. During ovipositing she always crossed her hind metatarsi behind the cell.   Sometimes she constructed two cells in different places at the same time, and then, while she laid her eggs in the one, the *lapidarius* workers, amid much quarrelling, would often seize the opportunity to lay their eggs in the other.   After she had finished guarding her own eggs she demolished the cell containing the *lapidarius* eggs.   She was heard to utter the mournful cry (p. 252) on many occasions.   On July 23 I removed her from the nest for four hours to let the workers finish laying eggs in a cell which she had begun and they had built up, but when I put her back the workers attacked her and stung her to death.   Although she had laid a batch of eggs only a quarter of an hour before she was removed, the workers did not molest them.   It may be the *Psithyrus* kneads a distasteful saliva into the wax covering her eggs.

On another occasion, a strong *lapidarius* nest in which a *rupestris* queen had been reigning for eleven days was overturned.   The shock of the accident seemed to unnerve the *Psithyrus*, for she wandered aimlessly about the nest, and two hours later the workers mobbed her and stung her to death.   It appears that when the colony is populous the *rupestris* will lose her life unless she maintains continuously her rule of repression.

On July 20 I put a searching *rupestris* queen into the vestibule of a *lapidarius* nest containing about 80 workers. She passed into the nest without hesitation and immediately produced an uproar, the workers and their queen rushing hither and thither in great excitement.   The excitement died down in about twenty minutes, and on lifting the nest I found the *Psithyrus* in a dying condition with six workers attached to her trying to sting her and thirteen dead workers lying around her.   A few days later I found in this nest a female of the deadly parasitic moth *Aphomia sociella*, and another of the deadly fly *Brachycoma devia*, but the humble-bees paid not the slightest attention to them.

Physocephala rufipes.—Mr. E. E. Austen of the British Museum has kindly examined the pupa that developed from a large dipterous larva that I found inside the abdomen of a queen of *B. terrestris*, stung to death on July 24, 1912, and has found it to belong to one of the Conopidæ, "very probably to the species known as *Physocephala rufipes*, Fabr., which is parasitic in the abdomen of the imago of *B. terrestris* as well as in that of other species of *Bombus*. It appears that when the parasite reaches the final larval or pupal stage the host frequently, if not always, perishes." The queen was one bred the previous year, and was taken with her nest, in which the first workers were about to hatch, on July 14. The day before she was killed I noticed that she was breathing at a slower rate than normal, and she had laid no eggs for at least two weeks.

INDEX

abnormal specimens, 149
affection of humble-bee for her brood, 30, 103, 116, 233, 269
agrorum, Bombus, 36, 42, 43, 44, 53, 68, 104, 148, **193**
antenna-cleaner, 10
antennæ, 8, 144, 145
  function of, 8, 62
  joints of, in males, 146
anger of queen, 228, 245
  of workers, 8, 51, 90, 199, 245, 254
Antherophagus, 78
Anthophora, 143
ants, 116, 129, 228
Apathus, see *Psithyrus*
Aphomia sociella, 73, 277
arcticus, variety of *Bombus agrorum,* 194
argillaceus, variety of *Bombus ruderatus,* 49, 177
armature, 144, 146, 222
  of each species, 271 (Plate VI.)
association between strange species, 57, 133, 136, 265, 275
auricle, 21-24, 171, 203

badgers, 81
barbutellus, Psithyrus, 69, 70, 72, **214**
beetles in humble-bees' nests, 18
Bingham, C. T., on the wing-hooks, 9
birds destructive to humble-bees, 81, 115
bluebells, experiment with, 267
Bombi, divisions of the British, 153
Bombus, derivation of name, 4
  life-history of, 12
Bombylius, 144
Brachycoma devia, 75, 127, 277
breathing apparatus, 11
breeding by selection, 142
  specimens, 223
Brightwen, Mrs. E., v
Brown-banded Carder-bee, 195

buzzing, how produced, 11
  to express anger and alarm, 8, 116, 138, 243, 252
  to express jealousy, 233, 235
  to express pleasure, 230

campestris, Psithyrus, 68, 70, 72, 147, **216**
canariensis, variety of *Bombus terrestris,* 159
Carder-bees, 17, 41, 152
caterpillars in humble-bees' nests, 73, 79
ceiling, waxen, of nests, 45, 243
centipedes, 120, 226
cheeks, 145
cleaning process, 268
clipping the queen's wings, 103, 266
coat, *see* hair.
cocoons, 29
  bitten down, 40
  first cluster of, 29, 33
  later clusters of, 35
  of different species, 36, 155, 159, 181, 185, 190, 199
  of *Psithyrus,* 69
collecting specimens, 220
colonies at night, 47, 99
  feeding, 123
colours, fading of, 148, 223
comb of humble-bee, 44
common carder-bee, 103
confinement, getting queens to breed in, 130
  keeping colonies in, 126
confusus, Bombus, 24
Conops, 78
corbicula, or pollen-basket, 20-24, 274
corsicus, variety of *Bombus ruderatus,* 178
cover, Sladen's, for domiciles, 109
covered way to surface nest, 47
Cryptophagus, 78
cullumanus, Bombus, 153, **171**

279

THE END

Printed by R. & R. CLARK, LIMITED, *Edinburgh.*

Printed in the United States
By Bookmasters